CAMBRIDGE LIBRARY COLLECTION

Books of enduring scholarly value

Earth Sciences

In the nineteenth century, geology emerged as a distinct academic discipline. It pointed the way towards the theory of evolution, as scientists including Gideon Mantell, Adam Sedgwick, Charles Lyell and Roderick Murchison began to use the evidence of minerals, rock formations and fossils to demonstrate that the earth was older by millions of years than the conventional, Bible-based wisdom had supposed. They argued convincingly that the climate, flora and fauna of the distant past could be deduced from geological evidence. Volcanic activity, the formation of mountains, and the action of glaciers and rivers, tides and ocean currents also became better understood. This series includes landmark publications by pioneers of the modern earth sciences, who advanced the scientific understanding of our planet and the processes by which it is constantly re-shaped.

Life of Sir Roderick I. Murchison

Sir Roderick Impey Murchison (1792–1871) was an influential Scottish geologist best known for his classification of Palaeozoic rocks into the Silurian system. After early military experience in the Peninsular War, he resigned his commission; a chance meeting with Sir Humphrey Davy led him subsequently to pursue a scientific career. *The Silurian System*, published in 1839, was a highly influential study, which established the oldest contemporary classification of fossil-bearing strata. Murchison was appointed President of the Royal Geographical Society in 1843. These volumes, first published in 1875, use information taken from Murchison's private journals and correspondence. Archibald Geikie (1835–1924) provides a detailed account of his mentor's life and work in the context of geology as a developing science in the early nineteenth century, and provides a fascinating insight into the life and work of this eminent Victorian geologist. Volume 1 describes Murchison's early life and geological studies until 1842.

Cambridge University Press has long been a pioneer in the reissuing of out-of-print titles from its own backlist, producing digital reprints of books that are still sought after by scholars and students but could not be reprinted economically using traditional technology. The Cambridge Library Collection extends this activity to a wider range of books which are still of importance to researchers and professionals, either for the source material they contain, or as landmarks in the history of their academic discipline.

Drawing from the world-renowned collections in the Cambridge University Library, and guided by the advice of experts in each subject area, Cambridge University Press is using state-of-the-art scanning machines in its own Printing House to capture the content of each book selected for inclusion. The files are processed to give a consistently clear, crisp image, and the books finished to the high quality standard for which the Press is recognised around the world. The latest print-on-demand technology ensures that the books will remain available indefinitely, and that orders for single or multiple copies can quickly be supplied.

The Cambridge Library Collection will bring back to life books of enduring scholarly value (including out-of-copyright works originally issued by other publishers) across a wide range of disciplines in the humanities and social sciences and in science and technology.

Life of
Sir Roderick I.
Murchison

Based on his Journals and Letters

VOLUME 1

ARCHIBALD GEIKIE

CAMBRIDGE
UNIVERSITY PRESS

CAMBRIDGE UNIVERSITY PRESS

Cambridge, New York, Melbourne, Madrid, Cape Town,
Singapore, São Paolo, Delhi, Tokyo, Mexico City

Published in the United States of America by Cambridge University Press, New York

www.cambridge.org
Information on this title: www.cambridge.org/9781108072342

This edition first published 1875
This digitally printed version 2011

ISBN 978-1-108-07234-2 Paperback

LIFE

OF

SIR RODERICK I. MURCHISON.

a

Rod.... Murchison

From a Painting by Pickersgill, R.A. Engraved by Joseph Brown.

JOHN MURRAY ALBERMARLE ST: 1875.

LIFE

OF

SIR RODERICK I. MURCHISON

BART.; K.C.B., F.R.S.; SOMETIME DIRECTOR-GENERAL OF THE GEOLOGICAL
SURVEY OF THE UNITED KINGDOM.

BASED ON HIS JOURNALS AND LETTERS

WITH NOTICES OF HIS SCIENTIFIC CONTEMPORARIES
AND A SKETCH OF THE RISE AND GROWTH OF
PALÆOZOIC GEOLOGY IN BRITAIN

BY ARCHIBALD GEIKIE, LL.D., F.R.S.

DIRECTOR OF H.M. GEOLOGICAL SURVEY OF SCOTLAND, AND MURCHISON PROFESSOR OF GEOLOGY
AND MINERALOGY IN THE UNIVERSITY OF EDINBURGH.

IN TWO VOLUMES—VOL. I.

Illustrated with Portraits and Woodcuts

LONDON
JOHN MURRAY, ALBEMARLE STREET
1875.

PREFACE.

COMPARED with foregoing periods of history, the nineteenth century has been marked by the extent and rapidity of its social transitions. These must undoubtedly be ascribed in great measure to the strides made by the physical sciences. Without claiming for Geology any prominent share in them, we may yet contend that this branch of science has done much to open out those wider views of nature and of man's place here, which have so powerfully influenced the tone and tendency of human thought and speculation at the present time. So that the history of a man who was a conspicuous actor in the drama of the establishment of Geology, as a science, may possess more than a merely individual interest.

The life of Sir Roderick Murchison was cast in this time of notable transition. Living on terms of intimacy with not a few of the leading men of his day, he himself bore a part in the leavening of the community with an appreciation of the nature and value of science. For many years he was in the habit of keeping a record of the events which he witnessed, or in which he took part. In the belief that the story of his life might have some interest and usefulness for those who should succeed him, he used now and then during his later years to devote his spare hours to the task of reading over his early journals, and superintending their transcription in whole or in abstract under his own eye. In the course of time a goodly series of closely-written volumes grew under the hand of the amanuensis, but their author at length perceived that their details could hardly possess sufficient interest for general readers. In the spring of 1871 he proposed to me that I should undertake the task of reducing his memoranda into a connected narrative.

Having accepted the office of biographer, I found that, in addition to the journals, there existed a vast mass of miscellaneous letters and papers going back even into last century. It appeared that Sir Roderick for many years of his life never destroyed any piece of writing addressed to him,—notes of invitation to dinner, and acceptances of invitations given by himself, being abundant among the papers.

To these materials, through the kindness of his friends and correspondents, to all of whom sincere thanks are due, I was subsequently enabled to add a large series of his own letters.

From the first it appeared likely that no narrative devoted merely to the personal events of Sir Roderick Murchison's life would be satisfactory. And as the work of arranging the voluminous materials proceeded, the desirability of adopting a wider treatment became increasingly evident. His life, closely bound up with the early progress of geology in this country, was one of work and movement. Duly to follow its stages, the surroundings among which it was passed must be constantly kept in view,—notably his comrades, their work, and its relation to his own. Accordingly I deemed it best, while keeping his story prominently before the reader, to give an outline of so much at least of these surroundings as would probably show with adequate distinctness what Murchison was, and what he did. With this view I have sketched some of the more salient features in the rise and growth of the geology of the older formations in Britain, including, at the same time, notices of Murchison's predecessors and contemporaries in the same branch of science. Obviously, however, even such a general outline as was alone admissible into a work like the present could not be continued into the later years when Murchison ceased to be the same prominent worker he had previously been, and when his labours were taken up and extended by others. To this historical aspect of the book, I believed that some additional interest might be given by a selection of portraits of some of the more conspicuous men to whom the establishment and spread of geology in Britain is due, more especially with reference to the study of the older rocks. Some difficulty was necessarily encountered in making the selection, arising in some cases from the want of available materials for the engraver, in others from the limited number of portraits admissible compared with that of the geologists deserving such recognition. Greenough, Fitton, and Lonsdale, for example, among the earlier luminaries, might have been most appropriately included in the list here given. To the friends who have supplied the paintings, drawings, and photographs from which this little gallery of scientific worthies has been engraved, my best acknowledgments are gladly given.

Of Murchison's early contemporaries who outlived him, and from whom assistance was received in the preparation of his biography, two of the most illustrious have since been removed by death.

Sir Charles Lyell furnished a series of letters on geological topics written to him by Murchison. Professor Phillips, besides supplying a large and most interesting collection of letters, which proved of great service in the preparation of the biography, kindly sent some memoranda of his own, which will be found incorporated in the book. To Mr. Poulett Scrope I am indebted for some interesting and useful notes respecting some of the older geologists of this country.

My friend and colleague, Professor A. C. Ramsay, has laid me under much obligation by the notes and suggestions sent by him as he read over the proof-sheets, and which are incorporated into the text or embodied here and there in footnotes. To Mr. John Murray, Mr. K. R. Murchison, Mr. Trenham Reeks, and Professor T. Rupert Jones, F.R.S., my thanks are likewise owing for a similar revision.

For the loan of letters written by Sir Roderick Murchison, acknowledgment is further due to Mr. Aveline, His Excellency Sir Henry Barkly, M. Barrande, Dr. Corbet, Lady Denison, Sir Charles Dilke, Sir Philip De Grey Malpas Egerton, Bart. ; Professor George Forbes, who supplied letters written to his father, Principal Forbes; Professor Johnstrup of Copenhagen, who sent a series of letters addressed to the late Professor Forchhammer; Captain Grant, Professor Harkness, Professor Hughes, who furnished the letters written to Sedgwick; Professor Hull, Major-General Sir Henry James, Mr. Martin, Mr. Hugh Miller, who procured a series of letters written to his father; Mr. K. R. Murchison, Mr. Murray, Mr. Lyon Playfair, C.B., M.P.; Professor Ramsay, Rev. Mr. Symonds, Mr. Todhunter, from whom came the letters addressed to Dr. Whewell.

CONTENTS OF VOL. I.

LIST OF ILLUSTRATIONS IN VOL. I.

CHAPTER I.

ANCESTRY—SCHOOL-DAYS.

AMONG the Western Highlands of Scotland there is no wilder tract than that which stretches between the Kyles of Skye and the line of the Great Glen. From the margin of the western sea the ground rises steeply into rugged mountains, which slope away eastward through many miles of rough moorland into the very heart of the country. The bold Atlantic front of these mountains is trenched by deep and narrow valleys, of which the upper parts rise above the sea-level into dark and rocky glens, the lower portions sinking under the water and forming the characteristic sea-lochs or fjords of that region. In the shelter of these hollows, alike in the glens, and as an irregular selvage along the margins of the lochs, lie strips of arable land with farm-houses and the cots of the peasantry; but all above and around are the wild rough hills, shrouded for great part of the year in mist, and catching the first dash of the fierce western rains, which seam their sides with foaming torrents.

Even now, with all the appliances of modern travel, these tracts of Lochalsh and Kintail are little known, except in so far as they can be seen from the sea, or from the few good roads which have been made through them. But some five or six generations back they were to all intents as remote from the civilisation even of the Scottish Lowlands as if they had lain in the heart of Russia. No roads led across them then. They could be traversed only by bridle-tracks, too little trodden to be always easily traced among the bogs and crags over which they lay. Notwithstanding the noble inlets which bring the tides of the Atlantic far into these wilds, there was then but little navigation, even of the simplest kind. Save the boats used in ferrying the lochs and in fishing, almost the only vessels ever seen were the smacks and cutters which from time to time smuggled ashore brandy and claret for the lairds.

Over this wild region the chiefs of the clan Mackenzie had for a long while held sway—a fierce and warlike race, exemplifying on their territory the curiously mingled merits and defects of the old Highland patriarchal system. In their midst, however, lay one or two smaller septs, sometimes in league with the dominant clan, sometimes in open arms on the side of their surrounding enemies. One of these septs went by the name of Mhurachaidh or Macmhurachaidh, that is, Murdoch or Murdochson, or, as it is now corrupted, Murchison. The first of the family must have been a Murdoch. Who he was, however, where he came from, and what he did to distinguish himself from the other abounding Murdochs of that part of Scotland, are questions to which no satisfactory answer seems now possible. Perhaps he was one of the Mackenzies, or more probably of the Mathiesons, or clan Malghamna, who possessed these tracts before the Mackenzies, and among whom Murdoch was a frequent name.[1] He may have been noted above his fellows for some characteristic, so that his posterity came to be called after him.

In the early part of the sixteenth century we find the Murchisons in possession of land in Kintail. In the year 1541, Evin M'Kynnane Murchison was proprietor of Bunchrew when he obtained a remission from James V. for having taken an active part, together with some of his neighbours, in burning the castle of Eilandonan, the stronghold of the Mackenzies, at the mouth of Loch Duich. It has been conjectured by a friendly genealogist, that for such deeds the sept received the soubriquet of " Chalmaon," or " brave ;" and that this title led to their being confounded with certain M'Colmans of Argyleshire.[2] There must at least have been a wonderful versatility about the race, for not many years after the raid on the Mackenzies, when the Reformation had already made way through the country, the churches of Kintail, Lochcarron, and Lochalsh were in peaceable possession of different members of the family.[3]

In the following century (1634) the Murchisons appear on the Ross-shire rent-roll as holding land in Lochalsh, of which they had obtained charters from the Crown. By this time,

[1] This suggestion has been made to me by Mr. W. F. Skene, who adds that "the small septs are often the remnants of the older population."

[2] In the North-West Highlands the Murchisons are called in Gaelic M'Colman, and have been traced by some genealogists to an origin in Argyleshire, where a sept of that name occurs. The family traditions, however, insist on a more northern origin, as stated in the text.

[3] In 1574, James VI. presented John Murchesoun " to the haill commoun kirk, baith parsonage and vicarage, of Kintail." In 1582 the same King presented Donald Murcheson to the same church, then vacant by the demission of John Murcheson, and Master Murdo Murcheson to the parsonage of Lochalsh and Lochcarron.—*Register of Great Seal.* For these references I am indebted to the courtesy of Mr. Skene.

too, they seem to have settled their differences with the
Mackenzies of Seaforth, for they then held rank as hereditary
castellans of that same Eilandonan stronghold which about
a hundred years earlier they had assisted to demolish.

It is not, however, until the troublous times of 1715 that
any member of the Murchison sept comes notably forward
in Highland history. Up till that period the people of these
wilds remained under the same clan-system which had
prevailed from the earliest times. The word of their chiefs
was their law, and they had but a feeble notion of any
higher rule or greater authority outside the dominions of the
clan. While this ancient obedience and attachment continued on the part of the vassals, the chiefs themselves were
more or less influenced by somewhat similar feelings towards
the old line of the Stuarts. A new race of sovereigns had
been installed by Southern and Saxon hands. It was regarded by these mountaineers with distrust and fear. They
had no great cause to look back with satisfaction to their
treatment under the sway of the fallen house. But there
appeared more risk than ever of molestation from the new
and alien rulers; and so, partly from loyalty to the Stuarts,
and partly from distrust of the Hanoverian dynasty, there
existed at this time among the Highlanders a wide-spread
disaffection and longing for a restoration.

At last these feelings found vent in open insurrec-
tion, and the outbreak of 1715 began. Among the chiefs
who appeared in arms came the Earl of Seaforth, head
of the Mackenzie clan. With him marched a gallant
company of Murchisons, including two of note, John and
Donald, uncle and nephew, the former bearing a commission
in the Prince's army, and bringing with him all the men he

could muster in Lochalsh, the latter holding rank as colonel, his commission having been sent over by the Pretender himself in a quaint large ivory " snuff-mull," inscribed with the words " JAMES REX.   FORWARD AND SPARE NOT." [1]

Among those who fell in the disastrous battle of Sheriff-muir was the great-grandfather of the subject of this biography.   Colonel Donald, however, made good his escape, and soon afterwards appeared in his native district, where, amid narrow inlets and bays, rough glens and lonely moors, he could bid defiance to the conquerors.

Donald Murchison was certainly one of the most remarkable Highlanders of his day.[2]   Bred a lawyer at Edinburgh, he united to the usual warlike virtues of the clansman a shrewdness and knowledge of the world, which gave him considerable influence as the agent and friend of the Earl of Seaforth.   After the battle of Sheriffmuir, when the Earl went into exile in France, Donald appears to have gone back to the mountains of Kintail.   Doubtless, in 1719, he took his share in the rude fortifying of Eilandonan Castle, of which, as we have seen, his family had been hereditary castellans, and saw with dismay its walls battered to pieces by the guns of three English war-vessels.   Nor was he likely to be absent from his chief when the luckless expedition of Spanish auxiliaries and Highlanders, marching eastward for the invasion of the country, encamped in Glen-

[1] This box was in the possession of Sir Roderick up to the time of his death, and is now one of the family heirlooms in the keeping of his nephew and heir, Mr. K. R. Murchison.   It forms a conspicuous feature in the picture of " Donald Murchison gathering Seaforth's rents in Kintail," painted for him by Sir Edwin Landseer, and bequeathed by him to the National Gallery at Edinburgh.

[2] For an account of him see Chambers's *Domestic Annals of Scotland*, vol. iii.

shiel.   Seaforth escaped wounded, and Donald was not among the prisoners.

The Seaforth estates were forfeited, but they lay in so remote and inaccessible a region that the Commissioners of the Forfeited Estates only in 1721 were able to procure a factor bold enough to march westward to take possession of them.   Donald Murchison, however, had been intrusted with their keeping by him whom he and all the native population still regarded as the rightful laird.   Hearing of the approach of the new factor with a body of the King's troops, he attacked them as they toiled through one of the savage glens of the district, and not only stopped their further progress, but compelled the factor to give a bond of £500 that he would never again attempt to carry out his duties in that quarter.   That he might have additional sanction for his own proceedings, Donald even extorted authority from the unfortunate official to act as deputy-factor for the Commissioners of Forfeited Estates, so that he could draw his rents for the Earl either as the agent of the one Government or of the other, as might be needful in each case.

Again, in the following year, a still larger party of soldiers made another attempt to gain possession of the rebellious country.   But once more Donald proved himself not unworthy of the colonel's commission and the ivory snuff-mull.   By a clever piece of strategy he discomfited this new invasion, and forced it to retire to its starting-place at Inverness.

For ten years Donald Murchison administered the Seaforth estates.   Even after his successful resistance to the royal troops, such was his boldness that he would go personally to Edinburgh to see after the proper transmission of the rents to the banished and attainted Earl. General Wade, in reporting to George I. in 1725, writes that "the rents [of the Seaforth lands] continue to be collected by one Donald Murchison, a servant of the late Earl's, who remits or carries the same to his master into France. . . . The last year this Murchison marched in a public manner to Edinburgh to remit £800 to France, and remained fourteen days there unmolested. I cannot omit observing to your Majesty that this national tenderness the subjects of North Britain have one for the other is a great encouragement for rebels and attainted persons to return home from their banishment." [1]

Though the "Coarnal," as Donald was called then, and as he still lives in old Ross-shire story, preserved the estates for the Seaforth family, risking often his life in the service of his master, the Earl, on regaining his position in his native country, treated his faithful ally with injustice and neglect. Taking advantage of the lawlessness of the time, he seized the charters and lands of the Murchisons. Donald, finding reparation hopeless, and despairing of success in any appeal to a Government which had no strong reason to be very active on behalf of a man who had given it so much trouble, retired to the east side of the island, and died of a broken heart, childless and in poverty.[2] He was buried by the Conon, but the memory of his deeds still lingers among the hills which he guarded so long and so well. Nearly a

---

[1] Wade's Report, in Appendix to Burt's *Letters*, 2d edit. (1822), ii. p. 280.

[2] For these particulars I am indebted to Dr. Corbet of Beauly, whose grandfather was a grandson of Colonel Donald's brother, and who has made the family genealogy a matter of investigation. See also Chambers, *op. cit.*, and Anderson's *Scottish Nation*, vol. iii. p. 731.

century and a half after he had passed away, a monument
was raised to him by his kinsman, Sir Roderick Murchison;
and now, as the tourist sails through the narrow Kyles of
Skye, and marks on one hand the mouldering barracks of
the Hanoverian soldiery, on the other the crumbling walls
of the castle of Eilandonan, a granite obelisk on one of the
headlands of Lochalsh recalls to him the deeds of one of
the most disinterested men of that wild time.

Donald's brother, Murdoch, raised an action at law for
recovery of the charters; but the renewed outbreak of 1745
came on. He took part in it, and died from the effects of
wounds received at Culloden. Thus the action disappeared,
and so did the ancestral property of the Murchisons.[1]

John Murchison, farmer at Auchtertyre, in Lochalsh, Sir
Roderick's great-grandfather, has been already referred to as
one of those who fell at Sheriffmuir. Traditions still linger
in the north as to his feats of strength; one large stone,
in particular, weighing about half a ton, being pointed out
as having been carried by him for some distance to form
part of a wall which he needed to build on his farm.

Of Alexander, grandfather of Sir Roderick, little has been
handed down. He continued to rent the farm of Auchtertyre,

[1] Sir Roderick was never able accurately to trace his relationship to
Colonel Donald. He seems to have regarded the hero as his great-grand-
uncle, but the connexion was yet more distant. His grandfather was a
third cousin of the Colonel, so that his own kinship was of that shadowy
kind in which Highland genealogists delight. Sir Roderick belonged thus
to an offshoot from the main stem of the Murchisons in whose hands the
little paternal property had been. His grandfather's great-grandfather
had owned it.—*Information from Dr. Corbet.*

Both Boswell and Dr. Johnson, in their narratives of their tour in the
Hebrides, refer with gratitude to the attention shown to them by a Mr.
Murchison, factor for the laird of Macleod, in Glenelg, who sent them a
bottle of rum, and an apology for not being able to entertain them in his
house.

and had to struggle with but slender means; yet, like his
predecessors who had not fallen in fight, he reached a good
old age, living on even till he was ninety-nine, and saw the
fortunes of the family retrieved by his eldest son, Kenneth,
whom he actually outlived.

It was in the year 1751 that this Kenneth came into
the world at Auchtertyre. He studied Medicine at the
Colleges of Glasgow and Edinburgh, took the diploma of
the Royal College of Surgeons in London, and while still
a young man went out as surgeon to India, where he
remained for seventeen years. A lucrative appointment
at Lucknow enabled him to amass a competent fortune,
with which, coming home again about the year 1786,
he not long afterwards purchased from his maternal uncle,
Mackenzie of Lentron, the small estate of Tarradale, in the
eastern part of the county of Ross. He appears to have
been a man of much force of character, a thorough Celt,
generous, yet with enough of worldly wisdom to keep him
from losing his possessions as his forefathers had done. He
wrote his journals in Gaelic, but used the Greek characters,
which he held to express the sound of his native tongue better
than Roman letters could do. Having gratified the ambition, so common in Scotland, to become a laird, he kept up old
Highland ways, and as long as he lived at Tarradale had as
one of his retainers a piper, who also played the harp. Fond
of antiquities, he devoted himself to those of Tarradale and
its neighbourhood, and made a collection of urns and other
objects found in tumuli and elsewhere on the estate. He
was one of the original members of the Highland Society of
London, and a warm friend of the scheme of the British
Fisheries for the employment of the people of the Western

Highlands and Islands. In those days doctors were scarce in the Highlands, hence Dr. Murchison's house formed a centre of attraction to the sick and maimed for many miles round. As he took no fees, his popularity became more wide-spread than was wholly pleasant, so that in the end he set on foot an agitation which resulted in the erection of the present Northern Infirmary at Inverness.

In the year 1791 he married the daughter of Mackenzie of Fairburn, lineal representative of the Rory More or Big Roderick Mackenzie to whom these estates had been granted by James v. She—as well as her brother, of whom more will be told in later pages—was born in the old tower of Fairburn, the characteristic Highland fortalice of the sept, guarding the entrance of one of the glens which open upon the lowlands of the Black Isle.

The first-fruits of this marriage appeared at Tarradale, on the 19th of February 1792, when the subject of this memoir saw the light. He received the name of Roderick, after his maternal grandfather, Roderick Mackenzie of Fairburn, a jolly old laird, who lived for more than ninety years, although, as he used to say of himself, in regard to whisky, claret, or other potations, he was " a perfect sandbank."[1] A second name was given to the boy—that of Impey, after Sir Elijah Impey, an intimate friend of his father's.[2]

For three years the family continued to reside at Tarra-

---

[1] This expression has been handed down by Sir Roderick Murchison. With reference to it Dr. Corbet informs me that he is himself in posses-sion of old Fairburn's silver quaich or drinking-cup, and that it does not hold more than an ordinary wine-glass. But of course the size of the cup tells us nothing as to how often it was replenished.

[2] In one of Sir Roderick's journals the following notice occurs bearing upon this period of his life :—" Old John Gladstone's wife was the dearest friend my poor mother had. She was a Miss Annie Robertson, daughter

TARRADALE, ROSS-SHIRE, THE BIRTHPLACE OF SIR RODERICK MURCHISON.

dale. This period, however, was too brief to fix any early
Highland impressions on the memory of the future geologist,
although he used afterwards to say that he ought to have his
Celtic proclivities fully developed, for he had been nursed
by the "sonsie" miller's wife of Tarradale, who hushed him
to sleep with Gaelic lullabies, and no doubt, after the fashion
of the country, gave him now and then, when he whimpered,
a taste of the famous whisky distilled on the adjoining
lands of Ferrintosh.

These three years of infancy formed the only prolonged
residence which Sir Roderick Murchison ever made in the
Highlands. His later visits were only for a few weeks at a
time in summer or autumn. That early stay at Tarradale
might have been indefinitely prolonged, so as to change the
whole tenor of his life, had his father's health continued good.
A delicacy, however, brought on probably by his Indian
experiences, induced Mr. Murchison to quit his northern
home for a milder residence in the south of England.

Among the earliest recollections which his son Roderick
retained was one dating from the time of this southward
migration. These were the days of highwaymen, and the
party had journeyed armed. The father, always anxious
that his son and heir should be a manly little fellow, presented one day a pistol at his head, bidding him stand fire.
His wife, fortunately, was sitting by and snatched away

of the Provost of Dingwall, Ross-shire. When my father married he proposed that the bride's great friend and bridesmaid should stay with them.
Finding that she was in very delicate health, he attended to all her ailments for a year or more, and when I was brought into the world, the first
young lady's lap on which I was dandled was that of the mother of the
present Chancellor of the Exchequer. She has often told me this herself,
and has expressed how much she owed to my father for his kind medical
attention."

the child, when the pistol, which was not supposed to be loaded, went off, and a volley of slugs passed through the window.

In a jotting found among his papers, and bearing date August 14th, 1854, the son thus recalls the memory of his parents :—" My father was a good violin-player, and had a fine Cremona, on which he brought out his native and Jacobite airs with much feeling; whilst my mother, dear soul, though never a skilful musician, played her reels on the harpsichord with so much point and zest that even now I can bring her full to my mind's eye whilst I was dancing my first Highland fling to the tunes of ' Caber Fey ' or ' Tulloch Gorum.' "

The change from Tarradale to the south of England did not avert the malady from which the invalid was suffering. He died in the year 1796. Of his closing days the following notes have been penned by his son :—" A recollection of him, doubtless often since brought to my memory by my dear mother, is that while my father was in the last stage of the disorder (liver-complaint and dropsy) of which he died, my little brother and self were sent from Bath to the then sequestered village of Bathampton, where he took leave of us. The opening of the red damask curtains of the lofty old-fashioned bed, the last kiss of my dying parent, and the form of the old-fashioned edifice to which the invalid had been removed, have been stereotyped in my mind."

On the death of her husband, Mrs. Murchison moved with her two boys to Edinburgh, where she took the house No. 26 George Street.[1] As soon as age allowed they were

---

[1] The younger son, Kenneth, became Governor of Singapore, and after-wards of Penang.

placed under the instructions of Bishop Sandford. Most of the Jacobites being either Catholics or Episcopalians, she found herself among friends in the small gathering which the disestablished Church could muster at that time in the metropolis of the north. Two years before his death her elder son revisited the little chapel near Charlotte Square to which his mother used to bring him. The lapse of more than seventy years had not wiped away the recollection of these early days, and he could yet recall how, one Sunday, their fat little cook Peggie, having incautiously ventured westward to her mistress's chapel, returned abruptly to the house, inveighing with indignation at the profanity of an organ, " for she cou'dna bide to hae the house o' God turned intil a playhouse."

The widow, still young and attractive, was not long in finding a second husband in Colonel Robert Macgregor Murray, one of the younger brothers of the Chief of the Macgregors. He, as well as his brothers, had been on intimate terms with Mr. Murchison in India, so much so that the Chief and his brother, Colonel Alexander, with Sir Elijah Impey, were left as guardians to the two boys.

The marriage of his mother broke up the home-life of young Roderick. Her husband was called to Ireland to aid in suppressing rebellion there, and as she determined to accompany him, it became necessary to place the boy, now in his seventh year, at school. Accordingly, in the year 1799, he was sent to the grammar-school of Durham.

More than half a century afterwards he spoke of the pang of the parting from his mother, and from Sally, the Dorsetshire lass, to whose tuition he used to attribute the English accent which he retained through life. Before leaving Edinburgh he could already read the newspapers with emphasis, and recite various pieces of verse.

But now a new and strange life opened out to him. At Durham he was domiciled, with some twenty other boys, in the house of one Wharton—a kindly man, who taught them French, and who, though himself a strict Catholic, never attempted to taint any of his pupils with a bias towards Popery.

Six years passed away at Durham. They could hardly be called years of study. The boy, indeed, toiled in some fashion into the sixth book of the Iliad, crossed the "pons" in Euclid, and picked up a little French, besides the ordinary rudiments of an English education. But the somewhat morose and severe manners of the head-master were not of a kind to make learning pleasant. Nor in the discipline of the school, stern enough in its way, and often aided from a bundle of hazel rods, was there check sufficient to control the waywardness of the wilder boys. Among these Roderick, or "Dick," as they called him, was always a ringleader. Breaking bounds was the least of his offences. Many an expedition did he lead against the town boys, and when not engaged in actual offensive warfare, he would be found drilling his school-fellows in military exercises.

Pranks, too, of the dare-devil kind were a favourite pastime. At one time he would be seen sitting on a projecting ornament or corner-spout of the highest tower of the Cathedral, to the horror of his comrades, who lay down in abject fright upon the "leads." He filled up more than the usual list of boyish escapades with gunpowder and on treacherous ice. The broken ground on which the

romantic old city of Durham stands lent itself eminently
to such feats.   There was one exploit which deserves a passing mention, since it was, perhaps, his earliest attempt to
explore what lies under ground.   Just beyond the archway
leading to the Prebends' Bridge lay the open mouth of a
drain which had its other end on the banks of the Wear,
some hundred yards below.   It had been a boast among the
boys to get down to the bottom of the vertical mouth.   But
"Dick" one day undertook to force his way down the whole
length of the conduit to its farther opening at the side of
the river.   Having dropt into his hole he soon found, as he
advanced on hands and knees, that to turn was impossible.
So, scaring many a rat by the way, he crept down, and at
last, with scratched skin and torn raiment, and probably
with what Trinculo styled "an ancient and fish-like smell,"
he emerged to the light of day, amid the hurrahs of his
expectant school-fellows.

His stepfather and his mother, during part of his stay at
Durham, rented Newton House, near Bedale, in the North
Riding, whither, in vacation-time, he repaired to exhaust himself in the delights of a pony and terriers.   There, too, it was
that the military life distinctly shaped itself in his mind.   His
maternal uncle, General Mackenzie of Fairburn, seeing his
active habits, told him that in due time he would make a
good soldier.   "From that day," he remarks, "I read and
thought of nothing but military heroes."

CHAPTER II.

THE six years' schooling at Durham, such as it was, formed all the connected general education which Murchison received, though he tried to supplement it after a fashion a couple of years later at Edinburgh. It was thought to be amply sufficient as a groundwork for the profession of a soldier; the more special training needed for the military life could be obtained elsewhere. Accordingly, in the year 1805, being now thirteen years of age, he was taken to the Military College of Great Marlow. Late in life he could recall how his stepfather sang amusing songs to cheer him on the way; how, on arriving in London, they " were quartered at the Spring Gardens coffeehouse;" and how surprised he was to see, " in the box next to us, gloating over his beefsteak and onions, the corpulent John, Duke of Norfolk."

At Marlow his aptitude for study was not more marked than it had been at Durham. His six books of Homer and the Latin which had been flogged into him were no help in aiding him to solve even simple questions in geometry and arithmetic. He was rejected, or, in the language of his comrades, " spun," and sent back to " mug," or study. " I could not do," he says, " the commonest things in geometry, and was a bad arithmetician—a foible which has remained with me."

When at length he had passed as a Cadet, he continued to introduce a fair amount of frolic among his not very arduous duties. C. 26—for that was his number in the third company—became as conspicuous a ringleader among the boisterous youths at Marlow as he had already been among the boys at Durham. He succeeded, however, at the same time, in acquiring some military habits, and a slender knowledge of tactics and drawing. He now, for the first time, had to learn subjects really interesting to him, and, as he had been formerly in the habit of drilling his school-fellows for mere amusement, it was now a congenial and not very difficult task to become a good drill-serjeant. From this time, too, dates the development of that singular faculty he had of grasping the main features of a district. His exercises in military drawing at Marlow first drew out this faculty, and led to the future rapidity and correctness of his " eye for a country," to which, in his scientific career, he owed so much.

As a reminiscence of these Marlow days he writes:—
" As each cadet cleaned his own shoes and belts, and blackballed his own cartridge-box, we really knew what a soldier ought to do. French polish was then unknown, and the blacking which we bought of old ' Drummer Cole' required much elbow-grease to bring out the shine ; so that I shall never forget, when the Duke of Kent (the father of our gracious Queen) reviewed us, how I admired his highly-

polished, well-made Hessian boots, and his tight-fitting white leather pantaloons."

Those who remember the veteran geologist in his later days, and recall the military bearing which marked him up to the last, will readily appreciate how strong an impress these Marlow days left upon him. While a cadet he was also somewhat of a dandy. He preserved memoranda of the names of the titled people he met when he paid a visit; how he delighted in the " smart curricle" of one distinguished acquaintance; how he rode " the well-conditioned hunters or chargers" of another; how he dined at a fine old mansion one day, and played at whist with the young aristocracy of the place the next. He had good opportunity for indulging these tastes during a visit which he paid in 1806 to his uncle, General Mackenzie, who was at that time commanding a militia force at Hull. And yet other qualities of his nature were also developing themselves. His uncle, who kept a diary, made the following entry on 29th January 1806 :—" This day my dear nephew Roderick left me. He is a charming boy, manly, sensible, generous, warm-hearted— in short, possessing every possible good attribute. I think he has also talents to make a figure in any profession. That which he has chosen is a soldier. He goes back to Marlow College on the 3d of next month."

The following year, at the age of fifteen, he was gazetted Ensign in the 36th regiment, but did no regimental duty for some time after his appointment. He writes of this epoch in his life :—" For the first six months after I became an officer I was supposed to be *completing my studies!* In reality I was amusing myself with all sorts of dissipation at Bath, where I passed my holidays driving ' tandems' and wearing clanking spurs.

" On leaving Marlow I was removed to Edinburgh, where my mother and relatives lived, and was placed in the house of Mr. Alexander Manners, the respected Librarian of the Faculty of Advocates, where I was associated with five or six other youths all older than myself. Having a recruiting party in the city under my orders, and with plenty of money to spend and balls to dance at, it may be well conceived that I did not gather together much knowledge. Still I picked up a few crumbs, which were destined to produce some fruit in after times. Unquestionably, this winter in Edinburgh materially influenced my future character. For example, I took lessons in French, Italian, German, and mathematics. I also attended a debating club, and wrote (such as they were) two essays on political subjects, of which of course I was profoundly ignorant. While the young powdered military fop (pig-tails and powder were then in the ascendant) affected to despise all dominies and philosophers, I could not be one of the table presided over by the bland and courteous old Manners without picking up many useful hints for future guidance."

Though he may have made some progress with his books at Edinburgh, he does not appear to have been quite as sure of his success in that way as he was of his mastery over the kicking horse in Leatham's riding-school. At the same time he took lessons in thrusting and parrying with the foil from an old French valet-de-chambre in the service of the Comte d'Artois, afterwards Charles x., who was then living in exile at Holyroodhouse. As the result of these various accomplishments he came to have such a good opinion of himself that when, at last, in the winter of 1807-8, he joined his regiment at the barracks of Cork, great was his

chagrin to find the officers very different from the high-
bred dandies he had expected them to be. They seem
to have been for the most part quiet, well-disciplined old
soldiers, who knew their work and did it, and who, moreover, had seen a good deal of active service on the Continent,
in India, and in South America. He was no longer the
important personage he had lately been with the " recruiting
party under his orders." But in a little while he discovered,
that what his comrades lacked in outward show they more
than countervailed in the best qualities of soldiers. He found
that the regiment had been a favourite with Sir Arthur
Wellesley in India. His messmates could tell many a story
of the cool daring of their old Colonel, Robert Burne ; how
he led his men at the storming of Seringapatam ; and how,
when at Buenos Ayres, the Spaniards had brought up eight
guns that completely enfiladed the road by which the British
force was retiring, he halted his brave fellows and said
quietly to them,—" Now, my lads, I 've come to lead you
once more to an assault. You see these guns ! If we don't
take and spike them our regiment will be swept away ;"
and then how he plucked a flower, and coolly placing it in
his button-hole, drew his sword, and in a quarter of an hour
had, with his grenadiers, spiked every gun and driven the
enemy back into the town.

Such tales vividly impressed the imagination of the
young Ensign. His ideal of a military hero had hitherto
been his handsome young uncle, General Mackenzie, in the
full blaze of martial uniform, and it was his ambition to
become the General's aide-de-camp. But he now came into
contact with a real tried hero, whom he thenceforth set up
as his model.

Colonel Burne was an excellent specimen of a type of officer now probably extinct. Cool and daring on the field of battle, he was a severe disciplinarian. His piercing dark-brown eye proved quick to detect a careless pig-tail, or a failure of pipe-clay either in gloves or breeches. He had drilled his men to the most perfect precision after the method then in vogue, insomuch that his had become what was called a " crack regiment" at the camp on the Curragh. But with all this attention to the laborious system of training which prevailed in his time, he knew how to unbend after his day's work was past. At the mess-table he would sit habitually from five till ten o'clock, setting an example to all his officers in the potation of port. He could not tolerate a drunken man, and he despised a young fledgling Ensign to whom illimitable draughts of his own favourite beverage proved in any way disastrous. He himself never showed any indication of being in the least degree affected, save that " his nose was gradually assuming that purple colour and bottle-shape which rendered him so conspicuous in the subsequent Peninsular war." Such was the brave and jovial leader whom the young Ensign of the 36th set before himself for imitation.

The regiment moved to Fermoy in the spring of 1808; but shortly thereafter a small army of about eight thousand men assembled at Cork for foreign service. Its destination remained secret, though it was shrewdly suspected to be designed for South America to retrieve there the honour of the British arms. The charge of it was given to Sir Arthur Wellesley, with General Mackenzie as his second in command. The latter resolved to take with him his nephew Roderick as an extra aide-de-camp. Such a post had been the dream of the young Ensign's life ever since he had entered on his military career, and it seems to have impressed him more each time he saw his uncle in all the pomp of command.

But the projects both of uncle and nephew were rudely broken. The unexpected successes of the rising of the people of Spain against their French invaders at once drew the attention of the British Government to that country. The expedition was ordered to proceed not to South America, but to Spain. With this change of destination came also an alteration in the command. General Mackenzie was not to accompany the force, and the expectant aide-de-camp had to bear his mortification as he best could.

But it was still his destiny to join the expedition, not on the Staff, but carrying the colours of the 36th, for in passing through Fermoy to take the command, Sir Arthur Wellesley left orders for that regiment to proceed to Cork within twenty-four hours. A hurried gathering of goods and chattels, a march of twenty miles, an inspection in the streets of Cork by Sir Arthur himself, and then a string of boats filled with the red-coats slipping down to the Cove and to the transports—thus suddenly the young soldier of but sixteen summers found himself face to face with the stern realities of war.

CHAPTER III.

SIX MONTHS OF THE PENINSULAR WAR.

BRITISH expeditions had come to hold but a poor reputation when the present century began. The despatch of a new one created little enthusiasm, or even interest. Long years of war had made the minds of men familiar with campaigns and battles and sieges. And these warlike operations were now spread over so wide a field that it would have been hard to tell to what quarter a fresh expedition would, with most probability, be sent. With this low military prestige there existed also a wide-spread feeling of indifference, sometimes bordering on contempt, for the profession of a soldier. The rank and file of the army contained a large infusion of the lowest orders of the community. Enlistment was in the hands of agents who received a profit according to the numbers they could induce to join the service. A man who had proved himself unfit for any honest calling was yet good enough for a soldier. And thus it became common to regard the " listing " of a son or brother as a kind of family disgrace.

Of the private himself but slender care was taken by the

authorities. He enlisted for life, and could look forward to being permitted to leave the service on a small pension only when ill-health or age at last made him useless. As a rule, he could neither read nor write. There was then no daily newspaper press recounting to every town and village in the three kingdoms the doings of his regiment, mentioning even his own name should he distinguish himself ; no associations for the help of the sick and wounded ; no lady-nurses venturing from dainty homes into the rough scenes of war; no frequent post bringing him letters and papers from the fatherland to show him that he was the object of kindly solicitude to his native country. When he was carried away into service abroad, it was not in a roomy steam-transport, but in a sloop or brig drawn perhaps from the coasting trade. And yet in spite of all these wants, of many of which he was happily unconscious, in spite, too, of pipe-clay and blackball, of plastering his queue, and burnishing his musket, he could be trained into an excellent soldier, and he went through his hardships with that endurance and boldness which more than restored the reputation of the British army.

On the 12th July 1808 the small expeditionary force set sail from Cork, and met with no mishap until it came to anchor off the coast of Gallicia. Owing to some uncertainty as to the state of affairs in the Peninsula, the disembarkation was delayed for a few days, and the transports moved southward to the Portuguese coast. The young ensign of the 36th regiment, cooped up in a small brig, had been in the surgeon's hands, and continued still an invalid. But at the order for landing his kit was soon packed. Like the other officers he took ashore three days' provisions, beside his greatcoat and knapsack, while he had to carry on his shoulders the colours of the regiment. Of this time he writes :—

"Early on the 1st of August, the 36th, forming part of the first brigade, disembarked with the 60th Rifles and other regiments under General Fane. Fortunately it was a fine calm hot day, with little or no surf on the sterile and uninhabited shore, with its wide beach and hillocks of blown sand. The inhabitants of Figuiera, on the opposite bank of the river, stood under their variously-coloured umbrellas, and my boat being to the extreme left, I could scan the motley group, in which monks and women predominated. Just as I was gazing around, and as our boat touched the sand, the Commodore's barge rapidly passed with our bright-eyed little General. Perhaps I am the only person now (1854) living, who saw the future Wellington place for the first time his foot on Lusitania, followed by his aide-de-camp, Fitzroy Somerset, afterwards Lord Raglan. He certainly was not twenty paces from me, and the cheerful confident expression of his countenance at that moment has ever remained impressed on my mind. The disembarkation being unopposed, you would think I had nothing to record. But the young ensign, with his glazed cocked hat, *square to the front*, his long white gloves, his tight belts, and well-filled knapsack and haversack, found it no easy matter to obey the orders of the fidgety General Fane, who, whilst our feet slipped back on the loose sand, was endeavouring to make us move as if on the Brighton race-course ! "

Of this toilsome march, and of the subsequent operations of the army, the young soldier wrote a minute and earnest

account two days after the battle of Vimieira, in a letter to
his uncle, General Mackenzie, which, with all its tediousness
of detail, shows no ordinary powers of observation, and grasp
of the general plan of the military proceedings :—

"VIMIEIRA, 23d *August* 1808.

"MY DEAR UNCLE,—Having been prevented so very long
a time from writing to you, on account of not knowing to
what part of the Mediterranean you are ordered, I am resolved at last to send this letter to Sicily, and let it run the
hazard of a ship sailing from Lisbon to that island.   If you
had been in England during the whole of the time in which
we were acting against the French in this country, what
pleasure it would have given me to have sent you from the
scenes of action the last accounts of them ; but in such
ignorance was I of the country you were in, that in the only
letter which I have had from my mother since I left Ireland, she informed me only of your having proceeded in the
' Pomona' frigate to the Mediterranean ; that it was probable
you would touch at some of the Spanish ports, whither it
was then supposed Sir Arthur Wellesley's expedition would
proceed ; and that in case of meeting with me, you intended
taking me on with you as your aide-de-camp.   I shall endeavour in this letter to give a detailed account of our proceedings, as I am certain you will be pleased with it, incorrect
as it may be in some respects, and far as it must be from
being a general one, on account of my humble situation in
the army.

"Sir Arthur Wellesley, after having proceeded to Corunna
in order to hear of the movements of the Spaniards, wrote to
Admiral Sir Charles Cotton off the Tagus, and requested him
to co-operate.   The landing of the troops in Mondego Bay

was then determined upon, and, on the 1st of August, the
36th and 40th infantry, and some rifles, disembarked on the
south side of the river Mondego, under General Fane, exactly
opposite the town of Figuiera.   The troops passed the bar
of the river chiefly in small schooners which trade along the
coast, and also in Portuguese boats.

" The brigade being formed was then marched in open
columns along the coast, chiefly through very heavy sands,
about two leagues, and encamped near the village of Lavaos,
where Sir Arthur established head-quarters for the night.
As by his orders two shirts and two pair of stockings and a
great-coat were to compose the whole of the baggage of
officers and soldiers, and that not such a thing as a donkey
or any other animal was procurable, our whole kit, including
three days' provisions, was on our backs, which, with a brace
of pistols and the 36th regimental colours, loaded me absolutely to the utmost of my strength.   Even our old Colonel
was compelled to tramp through the sands this day, which
he did with the greatest alacrity.   In four days the whole
of the troops and stores were landed without any loss.   As
we were now to wait at Lavaos for the arrival of General
Spencer's force from Cadiz, we had it in our power to communicate with the shipping, and I was thus enabled to land
my boat-cloak and a few other necessary articles, which
have since been of infinite use to me on outlying picquets
(under walls and without tents) and guards, and to buy a
donkey to carry them, which little animal is with me at
present.   In the course of three days General Spencer's force
arrived and immediately disembarked.   The army being then
arranged and divided into six brigades, we were placed under
General Ferguson with the 40th and 71st regiments.   The

appointment of this excellent officer (who, I think, is your
particular friend) gave us, the 36th, great satisfaction.

" Sir Arthur Wellesley's orders, previous to our landing,
were most explicitly and clearly written, particularly in
explaining to the troops the nature of the service they were
about to enter upon, and directing the greatest attention to
be paid to the religion and customs of the Portuguese. We
were likewise given to understand by these orders, that
through the whole of the war we should be *en bivouac,*
and no tents allowed for officers or men. On the 10th the
whole army directed its march to Leyria. It was intended
at first to have marched only three leagues, but upon information being received that a force had proceeded by the
sea-coast, in order to have surprised some of our outposts,
our march was continued until three o'clock next morning.
We then halted and took up our stations on a cold, bleak
moor, about two leagues from Leyria, having marched upwards of twenty English miles. Next morning we marched
to Leyria (where the inhabitants had been maltreated by
Loison), and halted on the south side of the city, whence I
went in to inspect it. There we were joined by the Portuguese army, which did not exceed in strength 3000 men.
From what I could observe, there were about four squadrons
of cavalry, good-looking, well-mounted dragoons, being the
garde de police of Lisbon, who had made their escape from
thence on hearing of our disembarkation. The Portuguese
infantry was in a most wretched state of discipline. On the
13th the army marched two and a half leagues, and halted
at Lucero, about a mile and a half on the south side of the
beautiful ancient abbey of Batalha, where the Portuguese
gained that celebrated victory over the Spaniards which

secured the independence of their country. At this place, for the first time, we got hold of a few straggling Frenchmen. Next day, the 14th, we proceeded to Alcobaça, and halted near it. The abbey is most magnificent, and delighted me more than any public building I have seen. The library and kitchen of the convent are well worthy of admiration. Part of the French army had just quitted this place.

" We had proceeded next morning about half-way between this town and Las Caldas ; when, approaching the small town of Albaferam, the French appeared in sight. Their army was drawn up in close column, and was ready for action. They however continued their retreat, and we advanced and halted near Las Caldas.

" Sir Arthur had received intelligence that the French General Laborde was strongly entrenched in the mountainous pass at the extremity of the valley in which the old Moorish fort of Obidos stands, and that General Loison was at no great distance from our right. The greatest part of the army was advanced from the valley to force the pass, while General Ferguson's brigade (with General Bowes's in its rear) was sent off to the mountains on the left, with the intention of cutting off Laborde's retreat. We were proceeding in this direction when the French appeared upon our flank, in consequence of which we formed line, and changing direction advanced, as the fog cleared, towards the enemy. We marched over about two leagues of hilly ground, and when within about one mile and a half of the pass we unexpectedly perceived the whole of the enemy in direct march to it, and immediately afterwards our riflemen opened their fire from the top of a hill upon one of the enemy's columns, who returned a volley and retreated a short distance.

" It fell to the lot of the Rifles, 5th, 9th, and 29th regiments, to force the pass, and to the last regiment especially, who, from the nature of the ground, could in some places only ascend up the hill in single files.   It was on this account that the 29th lost so many officers and men, including the gallant Colonel Lake, who was some paces in front of his regiment when he fell.   Just as we arrived at the foot of the mountains our artillery was brought into play, which no doubt annoyed the enemy's retreating columns, and three companies of our regiment were detached in order to support our light infantry, with the other light infantry of the brigade.   The enemy had moved off, however, from the shots of the rifles, and the distant fire of a few pieces of our artillery.   The 40th regiment was then detached from our brigade to cover the baggage, and as soon as the firing ceased we pursued our march through the pass.   Swiss and Frenchmen were lying dead on all sides.   As soon as we got through, General Ferguson's brigade, with the others which had not been much engaged, formed on a very extensive heath, and were advanced in front in order to charge the enemy if he would stand ; but Monsieur would only permit a few stray shots to be sent into his solid columns—he had received beating enough to satisfy him for one day.[1]

" On the 19th the army moved on to Vimieira, and passed over the very plateau on which we of the 36th were, two days afterwards, to have an opportunity of signalizing ourselves.

" The village of Vimieira is situated in a narrow valley, amid rising hills.   In our front, on to the south-east, is a wood upon a low eminence ; and in the rear, on towards the

[1] This was the engagement of Roliça or Roriça.

coast, are very high hills. On the summit of these hills, which lie exactly between Vimieira and the sea, the greatest part of the British army was posted. On a lower hill on the right, and a little in front of the town, was the Light Brigade, with the 20th regiment. This was an excellent post of observation. On the hill on the left was the 40th regiment, which was the left of our brigade, the 71st High-landers on their right, and the 36th being in the hollow exactly in the rear of the village. Close to our front was a small river. The position was rather more than two leagues from the sea. . . . We discovered some squad-rons and picquets of French dragoons. Several officers approached us, and one coming particularly near (I suppose he was sketching), Captain Mellish (General Ferguson's A.D.C.) offered the long odds to any one that, if permitted, he would dismount him.

"On the following morning, the 21st, about nine o'clock, the drums of the 40th regiment beat to arms. This was occasioned by their outlying picquet being attacked by some small party of the enemy which was greatly advanced. In ten minutes we were formed. Our brigade, led by General Ferguson, immediately crossed the little river and ascended to the hill on which we were about to fight. We had hardly commenced our uphill move before the advanced posts of our centre, in the hollow near Vimieira, on our right, com-menced a very heavy fire. We proceeded up the hill and formed line under its brow. A brigade of artillery was brought up with the greatest promptitude, and two guns, under Lieutenant Locke, being placed on the rising ground on our right, and the others on the left, three companies of the 36th were detached to the edge of the hill on our right, in order to protect the guns, which were soon annoying the advancing French close columns in the finest style with shrapnell shells, whilst our rifles and light infantry were firing in extended files as videttes.

" After some very hot and close work the centre of our army, at the village of Vimieira, repulsed the enemy.   There General Anstruther's brigade, with the 50th regiment, received the enemy in front of the village.   Colonel Taylor, who had charged with four troops, the only cavalry we had, viz., of the 20th Light Dragoons, was killed in a wood, whilst our heavy artillery, which was placed upon the hillock in front of the village, cut up the enemy most dreadfully.   The 50th charged them with the bayonet; the 43d met them in a narrow lane when in open column, and gallantly repulsed them; the 52d and 97th were likewise warmly engaged and thus the enemy was quite routed in their central or main attack.[1]

" To return to our own part of the battle, *i.e.* to our left wing: the fire of the enemy soon became very hot, and even though the 36th were lying on their breasts under the brow, our men were getting pretty much hit, whilst the regiment in our rear, the 82d, which at that time could not fire a shot, suffered more than we did.   General Spencer, who commanded the division, when moving about to regulate the general movements, was hit by a ball in the hand, and I saw him wrap his handkerchief round it and heard him say, ' It is only a scratch !'   Soon after, the light infantry in our front closed files and fell in; our guns were pulled back,

[1] The original of the present letter appears to have been lost.   In the copy of it from which the text has been printed, the remainder after the above paragraph is in Murchison's own handwriting of a much later date.

and then came the struggle. General Ferguson waving his
hat, up we rose, old Burne (our Colonel) crying out, as he
shook his yellow cane, that ' he would knock down any man
who fired a shot.'

" This made some merriment among the men, as tumbling
over was the fashion without the application of their Colonel's
cane.   ' Charge !' was the word, and at once we went over
the brow with a steady line of glittering steel, and with a
hearty hurrah, against six regiments in close column, with
six pieces of artillery, just in front of the 36th.   But not an
instant did the enemy stand against this most unexpected
sally within pistol- shot.   Off they went, and all their guns
were instantly taken, horses and all, and then left in our
rear, whilst we went on chasing the runaways for a mile and
a half, as hard as we could go, over the moor of Lourinhão.
They rallied, it is true, once or twice, particularly behind
some thick prickly-pear hedges and a hut or two on the
flat table-land ; but although their brave General Solignac
was always cantering to their front and animating them
against us, they at last fled precipitately, until they reached
a small hamlet, where, however, they did make a tolerable
stand.

" Here it was that Sir Arthur Wellesley overtook us
after a smart gallop. He had witnessed from a distance
our steady and successful charge, and our capture of the
guns, and he now saw how we were thrusting the French
out of this hamlet. Through the sound of the musketry,
and in the midst of much confusion, I heard a shrill voice
calling out, ' Where are the colours of the 36th ?' and I
turned round (my brother ensign, poor Peter Bone, having
just been knocked down), and looking up in Sir Arthur's

bright and confident face, said, ' Here they are, sir !'   Then
he shouted, ' Very well done, my boys !   Halt, halt—quite
enough !'

" The French were now at their last run, in spite of
every effort of Solignac to rally them.   Several of our
bloody-minded old soldiers said in levelling, ' they would
bring down the ——— on the white horse ;' and sure enough
the gallant fellow fell, just as the 71st Highlanders, who
were on our left, being moved round *en potence*, charged
down the hill, with their wounded piper playing on his
bum, and completed the rout of the enemy, taking General
Solignac of course prisoner.[1]

" Had we possessed a squadron or two of dragoons on
the left wing, all the remaining force of Solignac's division,
which had been driven two miles to the north, or away from
the main body of Junot (which had retreated to the south),
would have been captured, for they were then a rabble.
But Sir Arthur knew his weakness in cavalry.   He had
defeated a very superior force in crack style ; on our wing
we had indeed taken the General, and all the guns brought
against us; he also knew that the enemy had three full
regiments of cavalry in the field, whilst we had none.
Moreover, he was no longer commander, for old Sir Harry
Burrard, already on the ground, was his senior, and had
ordered a halt.

" Think, my dear uncle, with what pleasure I got a sheet
of long paper from the adjutant, and wrote my first account
of this glorious victory to my mother on a drum in the field,

---

[1] This appears to be a mistake.   Solignac was wounded, but the French
General taken prisoner was not he, but Brennier.   See Wellington's *Despatches*, vol. iv. p. 96 ; Napier's *Peninsular War*, vol. i. p. 215.

in order that it might go home with the despatches.[1]   We
shall soon go on to Lisbon, and then I expect we shall finish
off Monsieur Junot.—I remain ever, my dear uncle, your
most affectionate nephew."

To this letter may be added one or two reminiscences
which he used to tell of these first Peninsular days.   It was
no marvel if a stripling of sixteen, even though he had been
a ringleader in all rough sports and adventures at school
and military college, should have looked pale for a moment
on going into actual battle.   His face caught the eye of the
bluff old veteran, Captain Hubbard, who gave him a good
draught of Hollands gin out of his canteen, and patted him
on the back, saying he would never feel so afterwards.
" And he was quite right," added the narrator ; " the first
start over, and you are ever afterwards one of a united mass
of brave men."

No trace of personal emotion was of course allowed to
escape in the business-like letter to his uncle from the
embryo aide-de-camp.   And yet, brave and bold as he was,
he could not help a shudder at the first sight of the dead
and mangled bodies of the Swiss and French lying right
and left as his corps marched through the Pass of Roriça.
But a more hideous recollection dwelt in his memory
through life.   " When halting at a bivouac before we reached
Vimieira," he wrote, "a Portuguese volunteer on horseback
coolly unfolded before myself and others a large piece of
brown paper, in which he had carefully folded up like a
sandwich several pairs of *Frenchmen's ears*, his occupation
having been to follow us, and to cut off all these appendages

[1] This letter, sealed with a bit of brown bread, has not been preserved.

from men who were thoroughly well ' kilt '—doubtless to produce them in coffee-house in Lisbon as proofs of the number of the enemy he had slain ! "

The conduct of the 36th regiment, and its gallant colonel, received high praise in the despatches of Sir Arthur Wellesley, to which, in after life, Murchison referred with pride, as evidence that though his friends had almost all known him only as a civilian and a man of peace, he had yet had shared with his comrades in actual and successful fighting.[1]

The subsequent events of this short campaign, with all their memorable results in the Peninsula and at home, left but little impress on the young ensign. He saw his favourite general superseded by Sir Harry Burrard, and then by Sir Hew Dalrymple. He was quite sure that the British forces could have compelled Junot to surrender, or at least that the French force never could have fought its way back to Spain. Like so many of his fellow-countrymen, he looked on the so-called Convention of Cintra " as stupid, if not disgraceful." In spite of what he has described to his uncle as his " humble situation in the army," he seems to have had no hesitation in deciding that the brilliant successes in which he had taken part had been "shamefully lost " by subsequent diplomacy. And he no doubt found consolation in repeating

[1] In the official despatch from the field of Vimieira, Sir Arthur writes thus :—" In mentioning Colonel Burne and the 36th regiment upon this occasion, I cannot avoid adding that the regular and orderly conduct of this corps throughout the service, and their gallantry and discipline in action, have been conspicuous."—Wellington's *Despatches,* by Gurwood, vol. iv. p. 96. Again, in a letter written next day to Lord Castlereagh, he says, "You will see that I have mentioned Colonel Burne of the 36th regiment in a very particular manner ; and I assure you that there is nothing that will give me so much satisfaction as to learn that something has been done for this old and meritorious soldier. The 36th regiment are an example to this army."—*Op. cit.* p. 100.

to his comrades one or other of the contemporary squibs which expressed the popular estimation of the respective merits of the three commanders.

With the political side of the military events he troubled himself but little.   Of more interest at the moment were the sights of Lisbon, in which his regiment was now quartered, and the looks and ways of the inhabitants.   The music of the French bands before Junot's forces were embarked and sent away from the Tagus, the black-eyed beauties of the coffee-houses, and the filth of the luxurious city—these were the features of the sojourn in Lisbon which most impressed themselves on his memory.   Night after night his room was perfumed by the burning of lavender in it, and he was thereafter left to wage war against domestic battalions hardly less numerous than those which he had encountered at Vimieira.   Or if he ventured out of doors after nightfall, no little dexterity was needed to work his way safely among the discharges of filth, which, in accordance with the sanitary arrangements then in vogue, descended from the windows, too often followed, instead of being preceded, by the cry required by the police, of "Agua va!"

The month of September wore away.   At home fierce outcry had arisen over the Convention by which the French were removed from Portugal.   The three commanders and the leading generals were summoned back to England to undergo examination before a Court of Inquiry, while vehement denunciations were poured forth by the newspapers against the conduct of affairs after the battle of Vimieira.   Meanwhile events had transpired in Spain which wholly altered the aspect of the war, and gave occasion to the English Government to interfere more actively than ever in the contest between Napoleon and the people of the Peninsula. After the French armies had traversed Spain and crushed the numerous but unconnected and ill-directed attempts of the patriots to resist the march of the invaders, the tide of war turned. A division of Napoleon's armies, eighteen thousand in number, which had penetrated into the most southerly province, was surrounded by the insurgents and forced to lay down its arms. The enthusiasm of the people blazed forth afresh from one end of the country to the other. In England the joy was great and loudly expressed, that at last some check seemed likely to be placed on the career of conquest of the man whom the country hated and feared. Money, men, stores of every kind, were freely promised to the patriots, and as freely, though with sad want of judgment, supplied.

The British army, whatever might be thought or said as to the mode in which the feat had been accomplished, had certainly compelled the French to evacuate Portugal, and the Ministry of the day deemed it advisable that their victorious expedition, now lying at Lisbon and watching the embarkation and removal of the French regiments, should put itself in motion, march across the country, enter Spain, and give effectual aid to the efforts of the Spanish patriots. Orders to this effect reached Lisbon early in October. Sir John Moore was put at the head of the expeditionary force. He was told that not a French soldier remained in the southern half of Spain, that Castaños in the south, and Blake in the north, had collected large armies, with supplies, and how enthusiastically the people were everywhere rising against the invaders. He was directed to enter Gallicia or Leon, and there to receive an

additional force to be despatched under Sir David Baird
from England. In Spain his further movements were to be
regulated in concert with the Spanish generals.

Through the long melancholy marchings and counter-
marchings which began at Lisbon at the end of September,
and ended at Corunna in the middle of January, Murchison
took his place with the 36th. His regiment formed part of
the force sent round by Talavera under Sir John Hope. The
troops began to move as the rainy season was setting in. To
the rain succeeded the snows and frosts of an inclement
winter. From the Spaniards assistance neither in men nor
in means of transport, nor information of the movements
and strength of the common enemy, could be procured. To
the last there came from them in abundance promises of
powerful reinforcements, entreaties to the British commander
to advance, glowing pictures of the vast enthusiasm and resources of Spain, and stories of the weakness and hesitation
of the French. In the midst of so much uncertainty it was
natural enough that the progress of the British force should
be but slow, and that this tardiness and apparent hesitation,
combined with the hardships of the weather, should have
caused some murmuring in the ranks. Among the mur-
murers was our Ensign of the 36th. His physical frame,
though strong, was sorely tried during these long marches,
with indifferent food, in the dead of winter. He could not
then judge what were the real operations of the army. He
was necessarily ignorant, as other subalterns were, of the
almost incredible difficulties of the noble-hearted Moore.
He could see only the toilsome and seemingly staggering
marches and halts and retreats. It appeared as if at head-
quarters there were no settled plan; as if the army were

moved to and fro merely at random. So deeply was this
impression of inadequate generalship fixed on his mind, that
even late in life he continued to express himself as he might
have done in the march from Lugo, or on the heights of
Corunna.[1]

Of the actual events of the campaign he has preserved
notes, chiefly of the various stages reached by his division
in its march from Lisbon through Portugal and Spain, with
a few personal reminiscences. In a little pocket note-book,
which went with him through the campaign, there are traces

[1] The following note contains his deliberate judgment as to the generalship of Sir John Moore.   It was written about the year 1854 :—

"The chief mistakes of Moore can never, I think, be set aside, although,
doubtless, he had a most difficult task to play, and was grossly deceived
by the Spanish government.   These mistakes were, 1st, sending round all
his artillery and cavalry, when we entered Spain, by a long march, thus
paralysing his exertions for a fortnight or three weeks ; 2d, making the
hazardous and indecisive advance from Salamanca to Sahagun, which led
him eventually to abandon the only true strategical plan of returning, as
he himself intended a week before, on the strong ground of Portugal.
Again, the detaching the Light Division to Vigo was an error which prevented his occupying a strong position before Corunna ; and, lastly, his
forced night marches in order to escape from our enemy, who was re-
pelled by us at all points, even after our horrible losses and disasters, and
with two-thirds only of our army.

"It must be recollected that I only had the knowledge of a young
subaltern officer, and in resenting the stern general order of our chief, in
which he reflected on the want of discipline, I simply express what all the
poor sufferers felt who knew that the army so condemned was in an admirable state a month before. 'To whom therefore,' said we, 'is this
forlorn state due, but to the chief who commands us to do impossibilities—i.e., to march without shoes and provisions, and in dark winter
nights?'

"For these reasons, notwithstanding all the praise of his admirers, including William Napier, who had been drilled under him, I have never
been able to regard Moore as a first-rate general. As a general of division,
as a disciplinarian, and as a noble type of unblemished character and un-
flinching courage, he was without a rival. Peace be to his ashes ! and let
glory be ever associated with the name of the hero who in Egypt contributed so much to the success of Abercromby, and who, like his gallant
Scottish countryman, met his death in the arms of victory."

of some attempts to acquire a few words of Spanish. Such phrases as were likely to be of service in the march are carefully noted. He records how, having now been promoted to be Lieutenant, he made his first essay in horse-dealing,—an unfortunate adventure, by which he secured an animal whose legs, when seen by daylight, turned out to have been all duly pitched below the knee, and whose most sprightly movement consisted in rolling himself on the ground, his feet in the air, and his rider sprawling in the sand beside him, amid the laughter of the regiment.

From Abrantes to Castello de Vide he notes the broken features of the ground, which rises into heights crowned here and there with quaint old hill-forts, and sinks into fold after fold of cork-forest, with plenteous harbourage for the hairless black pig, "the best food in Portugal." Now and then during the halts he and a companion would sally out for the inspection of castle, forest, village, or town, as might happen. At the venerable fortress of Marvão, for example, scattering troops of black swine, he climbed up to the fortifications of what seemed to be a forgotten tenantless hold, when a challenge suddenly came from a ragged sentinel in dingy brown, and with a sorely rusted musket, dangerous only to the hands that might venture to fire it. The strangers were reported to the " Governor," and they found, as the whole garrison, a score of men yet more patched than the sentinel, with hardly a lock to any one of their guns.

The 36th regiment was the first of the division which crossed the frontier into Spain. He chronicles in the behaviour of the natives a strong contrast to that of the Portuguese. Though received with shouts of " Long live the English !—Long live King George !" he found the people cold and distrustful; and he speaks of the disheartening effect upon himself and comrades of the indifference and reserve with which the houses on which they were billeted were opened to them.

There was much in this march into the heart of Spain to arrest the notice of an observant eye—the forms of the great table-land, with its sierras and river-gorges—the antique towns and mouldering ruins going back even into Roman times—the ways and manners of the people. Of these various features no jottings occur in the journal, save only such scanty ones as to show that they were not passed wholly without notice. At the Escurial the force halted for six days. Many of the officers contrived during this interval to see Madrid. Murchison, being somewhat unwell, spent the time among the jolly brethren of the great gridiron convent. What seems to have made the most lasting impression on him were the large flasks of wine hung before the window of every cell to ripen for private use. But he retained a vivid recollection, too, of the splendours of the art collection, then still untouched by French spoliation, and of the solemn resting-place of the Kings of Spain.

It was while waiting at the Escurial for tidings of the Spanish forces, with which the British were to co-operate, that General Hope learned how utterly these forces had been routed and dispersed by the French, who, under Napoleon in person, were now rapidly approaching the capital. At once the route was changed, and by a skilful move the British division under Hope was united to the main body of the army led by Sir John Moore. In the course of this rapid march there occurred at the old Moorish city of Avila an incident, of which Murchison gives the

following account :—" Our poor fellows being well tired
were either asleep or dozing against the walls of the houses,
when they were roused by a tramping of horses' feet and
loud clashing of metal, sounding just like a cavalry-charge,
which caused a few to run for their arms, piled in the middle
of the dark street, whilst many more made a *sauve qui peut*
into the adjacent alleys.   The charge having cleared the
street, knocking down many a piled musket, our amusement was great to find that one old vicious mule, breaking
away from the muleteers, had carried with him a troop of
his associates, who came full gallop clattering down the
street, tossing our camp-kettles and all their burdens by the
way.   This was the enemy's cavalry that awoke us !"

Hard winter weather and a continued retreat began to
tell upon the discipline and the numbers of the British
troops.   On the 6th of January, on reaching Lugo, Sir John
Moore issued a general order, beginning,—" Generals and
commanding officers of corps must be as sensible as the
Commander-in-Chief of the complete disorganization of the
army."   Lieutenant Murchison, however, could see no signs
of any such disintegration in the 36th regiment at that time,
and it was only after the terrible night-marches which succeeded the halt at Lugo that his division merited in his eyes
the severe censure of the Commander-in-Chief.   These toilsome nights, with the constant pressure of the French, and
of even more resistless foes, bitter frost and snow, told, too,
upon his own strength.   On one occasion, after a fruitless
midnight march against the enemy, who was supposed to be
advancing to the attack, Murchison, commanding that night
an outlying picquet, threw himself into a corner of a farmer's
yard, and soon fell asleep.   Day had scarcely broken when

the cry of "Picquet, turn out!" roused him from his rest, but not in time to escape the notice of the vigilant Colonel Packe, who, however, allowed him to escape with a severe reprimand. But after the halt at Lugo, when having vainly offered battle to the French, the British army retreated by a forced march to Corunna, the young lieutenant fairly broke down. The mule, which had hitherto carried himself or his kit, was lost; his old soldier servant had gone back to seek among the snow for his wife and child. Of this sad time he has preserved the following recollections :—

"Never shall I forget the night which followed the abandoning of our position in front of Lugo. We marched through that city at dusk, and then blew up the bridge which was to check for awhile our foe. In darkness, with no food, and after sleepless nights, with worn-out shoes, and thoroughly disgusted with always running off and not fighting, this army now fell into utter disorder. Starved as they were, the men soon became reckless, and all the regiments got mixed together; in short, the soldiers were desperate, in spite of the exertions of the few mounted officers. For my own part, I walked on, usually in my sleep, with the grumbling and tumultuous mass, until awakened by the loss of my boots in one of the numerous deep cuts across the road, which were like quagmires, so that with my bare feet I had some twenty miles still to march. Many of the soldiers got away from the road to right and left. Marching all that dreadful night my young frame at last gave way, the more so as I was barefoot, cold, and starved, and already the great body of troops had got far ahead of me. In short, I was now one of a huge arrear of stragglers when day broke, and the little hamlet was in sight.

" Seated on a bank on the side of the road, and munching a raw turnip which I had gathered from the adjacent field, and just as I was feeling that I never could regain my regiment, and must be taken prisoner, a black-eyed drummer of the 96th came by from the village, whither the young fellow had been to cater. Seeing that I was exhausted, and almost as young as himself, and not yet a hardened old soldier, he slipped round his canteen, which he had contrived to fill with red wine, and gave me a hearty drink. He thus saved me from being taken prisoner by the French, who were rapidly advancing, and who, if they had had a regiment of cavalry in pursuit, might at that moment have taken prisoners, or driven into the mountains, a good third of the British forces.

" With the draught of wine I trudged on again, and came in, at eleven o'clock of the 10th, into the town of Betanzos, and rejoined my regiment, which had marched in with about fifty men only, with the colours, though ere night it was made up to its strength of 600 and odd men. This fact alone shows better than a world of other evidence what forced night-marches with a starving and retreating army must infallibly produce. At Lugo the 36th regiment was fit to fight anything—in two days it was a rabble.

" Happily for me I tumbled into a shoemaker's house. His handsome young wife washed my feet with warm water, and furnished me with stockings, while her husband came to my further aid with shoes. But my swollen feet had no time to recover. On the following day the whole army, such as it was, passed over the river, blowing up the bridge, and taking up its last position.

" There, remnant as it was, the army formed a respectable

line—Corunna within two miles of us, and our fleet ready to
back us.   Provisions and shoes were served out to us, and
with such luxuries the bivouac, even in the month of January,
was well borne.   In truth the army got into comparative
good spirits, and when on the 15th the French crossed the
last bridge we had blown up, and were defiling at a respectable distance along our front, we were quite refreshed, and
ready to repel them.   The picquets indeed of our (Hope's)
division had a sharp encounter on that evening, and when
looking through the Colonel's glass, I saw Colonel Mackenzie
of the 5th regiment fall dead from his grey horse, whilst
leading an attack on two of the enemy's guns.

" On the 16th, just after our frugal repast, and whilst
leaning over one of the walls where we lay, my old Colonel
after looking some time with his glass, suddenly exclaimed
to me, ' Now, my boy, they 're coming on ;' and when I took
a peep to the hills beyond on the right and south-west, I
perceived the glitter of columns coming out of a wood.   And
scarcely had the Colonel given the word to fall in, when a
tremendous fire opened from a battery of seventeen to twenty
pieces, under cover of which the enemy was rolling down in
dense columns from the wooded hills upon our poor fellows,
who were in a hollow with their arms piled, like our own,
until they were assaulted.

" For our cavalry was extinct, as the horses and men, as
well as most of our artillery, were embarked on the 13th and
14th ; yet never since Englishmen fought was there a more
gallant fight than was made by the 4th, 42d, and 50th regiments (Lord W. Bentinck's brigade), who rushed on with
the bayonet, and, supported by the Guards, held their own
against a terrific superiority, until General Paget was ordered

to move his brigade towards the enemy's flank, and compelled them to withdraw—not, however, before poor Moore, galloping out from the town, fell, while encouraging the troops; and Baird, who marched his division out of the town, had lost his arm. My own brigade had much less to do, our front line and picquets being alone engaged.

" As night fell, and after the firing had ceased, the enemy having returned to his own ground, we received the order to march into Corunna and embark. Our fires were left burning to deceive the enemy, and make him believe that he must fight us again next morning if he hoped to beat us.

" Silently and regularly we moved on on this our last short night-march in the dark tranquil night of the 16th, and passing through the gates reached the quay. The names of our respective transports had previously been explained to us, my own being the brig Reward,' which I found to be from Sunderland. I was on deck as light dawned, and then at once saw the danger of the position of this miserable little transport, as well as of a dozen or more of the same craft. They had been foolishly allowed to anchor immediately under the tongue of high land which forms the eastern side of the harbour, and on which there were no land defences. Knowing that this ground was only a continuation of the hilly track on which my division had marched a few hours before, and being certain that the French would with the peep of day pass over our old bivouac to this promontory, I at once urged our skipper to get up his anchor betimes. But the grog had, I suppose, been strong that night. He exclaimed, 'Why, I tell you what, the brave Highlanders are there ; they have not come away like you folks.'

Scarcely had he spoken when a battery of field-pieces opened their fire and sent some balls through our rigging. Turning pale as death under the fire of these mere field-pieces, and seeing that his crew were ready to run below, he applied the axe to the cable, and in a few minutes we were drifting away as we best could. The wind being from the east, we were fast approaching the rocks on which the Castle of Antonio stands, and on which at least five transports similarly circumstanced to my own were wrecked, the men being saved with difficulty, after losing their arms, colours, and baggage.

" I have often reflected on the extraordinary want of all due arrangement on the part of our Admiral, in command of a splendid fleet, who allowed those miserable transports to anchor in such a position without placing a frigate or two near them to silence the puny battery and prevent the dismay which seized the skippers.

" Not 'missing stays,' the ' Reward' floated away, and was soon going fast before a strong nor'-easter, with the rest of the fleet helter-skelter for the Channel. The retreat from Lugo could not be more confused than this flight of ships. On the night after our start I was awakened by a strange noise, and running on deck found the ship wearing off under a furious storm from amidst white foam and breakers. We had just avoided going ashore upon the Dodman—a headland of Cornwall—which that very night sent three or four of our careless transports to the bottom with their crews, and filled with poor soldiers who had escaped from the dangers and privations of the campaign. Such were our transports of the old war. We had been saved from this disaster solely by the watchfulness of an old grenadier."

So ended Murchison's first and last campaign. After the lapse of more than half a century spent in peaceful and utterly different pursuits, and when men had ceased to think of him as having tried in any degree the rough ways of war, he loved to recall those old Peninsular days. Many a time did the recollection of them furnish him with a telling point in an after-dinner speech, and give to some of his hearers a surprise when they learnt that the speaker whom they had known or heard of, perhaps from boyhood, only as a man of science, had fought with Wellesley and Moore before the year of Waterloo.

From the end of January 1809 to nearly the end of October in the same year, Murchison remained with his regiment on home service, continuing to vary the routine of garrison life by visits to different parts of the country, among others to Tarradale, the paternal estate in Ross-shire. London, too, lay so temptingly near to Horsham Barracks, that he was often to be found with some of his messmates at the Old Slaughter's Coffee-house, St. Martin's Lane, then a favourite military haunt. On one of these occasions, escorted by his commanding officer, Colonel Burne, he was parading Bond Street in the stream of fashionable loungers when Sir Arthur Wellesley came up. The hero of Vimieira had for the nonce turned his sword into the pen of the Chief Secretary for Ireland, and his military uniform into a civilian's garb so unique that it remained ever after in the young lieutenant's memory :—" Coat double-breasted, with brass buttons, buff waistcoat, kerseymere shorts, and brown top-boots, leaving a good deal of daylight behind." Recognising the Colonel, he stopped. His words not less than his dress made one of the reminiscences which Murchison liked most

to recall. " Ah, my dear Burne," said he, "glad to see you
once more. One of your younkers—eh? Well, things won't
do as they are. I shall soon be at it again, and then I can't
do without the 36th." But though this prophecy came true
enough, and though doubtless the subaltern went away rejoicing in the prospect of again having a chance of distinguishing himself, he was not destined to take any part with
his regiment in the brilliant adventures which ended with
Waterloo.

Curiously enough, the very advancement which he had all
along contemplated as the height of military bliss became
in the end the ruin of his professional prospects. He now
attained his ambition, for in the autumn of 1809 he became
aide-de-camp to his uncle. But the change, though it led
him abroad, brought him no opportunity of advancing himself in his career.

General Mackenzie was then in Sicily, and his nephew
had orders to join him there. On the 25th of October,
George III.'s jubilee, he set sail. As the ' Salcette' frigate,
in which he had obtained a berth, slipped round the North
Foreland and down the Channel, the shores of Kent
from headland to headland, and from tower to tower,
blazed with cannon, while a great fleet fronting the coast-line
answered with one long flame of fire from ship to ship, as if
to show not merely loyalty to the old King, but a front of
defiance to be seen and understood by Napoleon on the
other side of the strait.

Life abroad wore now a pleasanter aspect than it had done
for him in the Peninsula. " At Messina," he says, "I was
soon set up as my uncle's aide-de-camp in a house of my
own, with two horses, and little to do except make love and

ride in the cool of the evening with my general." As one
of his duties he had to copy an official correspondence between his uncle and the agents of the Neapolitan Government, and thereby had an early opportunity of learning something of the duplicity and broken faith with which the
British in Sicily had to deal. Another correspondence also
copied out by him was one with Admiral Collingwood, then
in command of the Mediterranean squadron, whose despatches were pointed out to him by his uncle as
models for imitation.

A lull had come in the warlike operations in Italy. The
hostile forces, looking at each other across the narrow Strait
of Messina, contented themselves with a wearisome and
profitless gun-boat bombardment. Murat came down into
Calabria, and threats were given out that he would invade
Sicily and call on the people to rise against the hated
Bourbon ; but as no such move was made, the bombardment
went on.

This uninteresting duel was once enlivened by an incident worthy of an older time. A flag of truce came sailing
across from the French lines, and keen grew the interest on
the Sicilian side to learn what new turn affairs had taken.
Still greater, however, was the astonishment of everybody
when the French officer, disembarking with a package under
his arm, made known his mission thus :—" Le Roi mon
maître ayant appris que son bon ami le Général Mackenzie
se trouve en face, désire renouveler leur amitié, et lui envoye
quelques livres de bon tabac de Paris !"

It turned out that some years before, Mackenzie had
obtained leave of absence to go from Minorca to visit Rome.
While he was in the imperial city, the French army under

Murat suddenly appeared. The young British brigadier resolved not to flee, like most of his fellow-countrymen, but to trust to the effects of a bold bearing upon the generous and susceptible mind of Murat. On the evening of the French entry into Rome, a Princess, with whom Mackenzie was acquainted, gave a grand ball, at which he was announced in full uniform as "The English General." Taking no notice of the French officers, who looked at each other in astonishment, he saluted the hostess, and had entered into conversation with her, when at last Murat, recovering from his surprise, tapped him on the shoulder, and begged for some explanation. Mackenzie easily satisfied him that he was what he pretended to be,—a young British officer, "fond of pictures, pretty women, and amusement; and that as he was simply amusing himself and learning Italian, he thought he had better trust to the generosity of a brave General-in-Chief than be captured by troops and treated as a spy." Murat not only granted him leave to stay in Rome, but gave him a passport to travel where he pleased, and formed a friendship which was now renewed even in the midst of actual war.

As a further reminiscence of this friendship, his nephew writes,—" When the General [Mackenzie] visited Paris at the peace of Amiens, he found in Murat a most useful and kind friend, who presented him to the First Consul, with whom he dined. It was my uncle's habit to eat slowly, and in short to dine like a gentleman, in conversing with his neighbours. Massena, who was next him, said,— ' Dépêchez-vous, mon Général—le dîner sera bientôt fini et vous n'aurez rien à manger.' Such was Bonaparte's rapid and voracious mode of feeding (no wonder he died of a

cancer in his stomach !), that before my worthy uncle had
eaten the second dish, Napoleon was trotting by him, followed by all his clattering suite, to have coffee in the next
room of the Tuileries."

Although actual warfare was going on within sight of
Messina, our young aide-de-camp began again to complain
of monotony. He took pains to acquire some knowledge of
Italian, and, what may surprise those who knew him only
late in life, had lessons in singing. Of professional work
there would seem to have been but little for him to do ;
hence the arrival of a stranger, who needed to be taken
round the outskirts of Messina, was no doubt a welcome
excitement. His journals contain jottings of such short
excursions, parties, and other gossip. The only incident
beyond the usual routine relates to an English lady, one of
the beauties of the place, who, however, had the misfortune
to be extremely stout :—" One day at the table of the Commander-in-Chief, the captain of a Turkish frigate being
seated opposite to F—, was so lost in admiration of her
that D— and myself, who were sitting on either side of
him, asked him how much he would pay for her, and he
instantly replied, with sparkling eyes, 'Fifty brass cannon,'
—in other words, his frigate's worth."

General Mackenzie's health now required his return to
England, and our aide-de-camp was soon relegated once
more to home life. The journey homeward proved more
circuitous and prolonged, as well as somewhat more eventful, than the voyage out had been. They had berths on
board a " miserable little packet, with some six pop-guns,"
and their route lay by Malta and Cagliari to Gibraltar. Off
the coast of Sicily they ran a narrow chance of being sunk

by an Algerine squadron, the Algerines being then at war with the Sicilians. At Cagliari they beheld his Sardinian Majesty drawn down one of the steep streets of the place in a rickety coach by four black long-tailed horses. Ten days passed pleasantly away at Gibraltar, enlivened by an excursion into the hills of Ronda, in the wake of the retreating French, with the risk of being taken prisoners by them, or of being shot as Frenchmen by the guerillas. At Cadiz he made fresh acquaintances, witnessed a little further warfare in the attack and defence of Fort Matagorda, and enjoyed for a fortnight the evening stroll on the Alameda. The packet direct from Constantinople to England took him finally home.

CHAPTER IV.

THE military career of Lieutenant Murchison had now come wholly to depend for its shaping upon that of Lieut.-General Mackenzie. As the latter on his return home was appointed to command in the north of Ireland, his prospects of future advancement suffered hopeless ruin, and with them went those of his young aide-de-camp. Both aspirants for distinction were doomed to inaction at home just as Wellington was beginning his brilliant successes in the Peninsula, and they remained here through those eventful years— 1811 to 1814—during which the British army established its prestige on the continent of Europe.

With this forced inaction Murchison used to connect an incident illustrative of one phase of the society of England at the time. General Mackenzie had been a favourite with the Prince Regent, and continued to be so until one fatal night after his return from Sicily. The story is thus told by his nephew :—"My uncle was in the pit of the Opera when Sir A. Murray, the gentleman-usher of the Princess of Wales, came down to him from her box and conveyed the flattering

message that her Royal Highness wished to see him.   Hesitating for a moment, for he well knew how the Prince hated
her, he unfortunately assented, in the belief that no one
could refuse a royal command.   Of course, the Princess
having got one of the Prince's clique, and a handsome
fellow, in hand, made the most of her conquest, not only by
parading him in front of the box, but also by taking him
home to sup with her.   The late Lord Hertford, who was
the constant gossip of the Prince, went at the usual hour
next morning, and whilst H. R. H. was shaving said,—
'Well, Sir, strange things come to pass.   Mac was
with the Princess in her box last night, and went home
with her to supper.'   The razor fell from the royal hand,
and at once he took a dislike to my uncle, who never saw
him afterwards.   But to soften his fall the Grand Cross of
the Hanoverian Order was sent to him, the Prince saying,
'Mac is a handsome fellow, and will look well in it.'"

On his return from Messina, Murchison had again to
betake himself to dull barrack-duty at Horsham with the
second battalion of the 36th regiment, to which he belonged.
He had not yet discovered any form of mental occupation
which might serve to make even that monotonous sort of
life not unprofitable.   On his own confession, he gave himself up to walking feats, lessons in pugilism, horses, and the
other pursuits with which a young military dandy contrives
to fill up his time.   In the midst of this aimless life he
gladly obeyed a summons from his uncle to join him as
aide-de-camp in the north of Ireland, where the General
had been appointed to the command of a division.

Everything at first promised well in this new sphere of
action.   But when he had fairly settled down in his quarters

in the town of Armagh, the aide-de-camp found them even
more intolerably dull than Horsham, with a vastly greater
distance from anything like the pleasures of society. His
companion at this time, the Comte de Clermont, a young
French *émigré*, holding the rank of captain in our service,
had been appointed by General Mackenzie to be aide-
de-camp with Murchison. With every disposition to be
amused, the two young men found it no easy task to keep
themselves in good humour in Armagh. Having no kind
of military duty to perform, they spent their mornings in
hare-hunting with slow beagles. During the day they were
often to be found at a neighbouring rectory, drawn partly
by the whimsicality of the jolly parson, and partly by the
charms of his young ladies, among whom each of them contrived to fall deeply in love. From the rector's humour and
Miss B—'s attractions the change to the dull lonely evenings
at Armagh was no doubt intolerable. Now and then a tea-
party came off in their honour. When that form of excitement failed they had the chance of a game at tric-trac with
the General, who however would dismiss them at nine
o'clock to their lodging over a bootmaker's shop.[1]

In his journal of this period there occur allusions to the
Cathedral Library, but he appears to have made little use of
it, his chief mental exertions having been given to the discipline of his stable and the doctoring of his horses. Such
reading as he accomplished seems to have consisted of

---

[1] An amusing glimpse into this Armagh life is furnished by the
remark of a French cook whom the General had taken over to Ireland
with him, and whose disgust with the want of resources for his art, and
the intolerably pungent peat-smoke, found vent at last in the following
words, duly chronicled by the nephew :—" C'est avec infiniment de regret,
M. le Général, que je vous quitte : mais en vérité si je reste ici je perdrai
et ma réputation et ma vue."

Shakespeare and any sensational form of literature which
came to hand.

At this time of his life Murchison was simply one of
those numerous young men who, finding in the routine of
their military duty occupation for but a small portion of the
day, and having little inclination for pursuits requiring any
degree of thought, yet happy in the possession of excellent
health, strong bodies, and good spirits, need to get an outlet
somehow for their superfluous energy. Nor does he seem
to have been more fastidious than others in his choice as to
the direction in which that outlet was to be sought—feats
of pedestrianism, hunting, or horsemanship offered a ready
relief from the tedium of military idleness.

Now and again he obtained leave to go to England, and
on these occasions, when not following the hounds in the
northern counties, he was usually to be seen dressed in the
height of fashion and airing himself on the promenades of
London. For he had now managed to pick up expensive
tastes, and indulged in an extravagance which brought him
a series of earnest expostulations both from his guardian and
his uncle. On his own confession he spent treble and
quadruple his allowance, and looked forward to his majority
as an event which would enable him to gratify even more
freely his fondness for display. He even talked of selling
the patrimony in Ross-shire so soon as it came into his possession—a purpose which his guardian contemplated with
horror as a frustration of the design of Dr. Murchison, who
had purchased the property as an investment for the family,
and who held that a small freehold estate gives a man a
better position in the country than treble its value in the
bank. Murchison of Tarradale would have a voice in his
county, Murchison of the funds could have none.

In the midst of this purposeless extravagance it is plea-
sant to find a glimpse of better things. On the 27th of
January 1812, Captain Murchison became a Member of the
Royal Institution, where he attended the lectures of Sir
Humphry Davy.   No notice of this part of his London
doings, however, occurs in his journals.

At last the long-wished-for 19th February 1813 arrived,
and the young laird came of age. His guardian had urged
him to go north, see the property with his mature eyes,
and judge for himself whether he would act wisely in parting
with it.   He now resolved to follow this advice.   In those
days it was common to make the journey into Scotland on
horseback, or to post in one's carriage.   Young Tarradale
combined the two kinds of locomotion, for he converted
his tall hunter " Buckran " into a buggy horse, and with his
groom " started off steadily in his high green dog-cart."
After a short stay in Edinburgh he took the old Highland
road, and had reached Blair-Athol by the last day in March.
Next morning a loud thumping at his bedroom door, and
the voice of his Yorkshire groom—" Sir, I canna get in to
Buckran ; the snaw's blocked oop t' way to steable," brought
before him in a way not to be forgotten one of the risks of
Highland travelling in the old days.   Half a century afterwards he was again driving with the writer of these lines
along the same road, and recalled the picture of his escape
how after incredible labours, and with a strong gillie or two at
each wheel, he managed to reach the little wayside inn of
Dalnacardoch ; how the stage-coach, trying to follow them
late in the day, was capsized over the bank of the Garry,
and the driver, guard, and passengers, after trudging for some
miles through the snow, arrived with nightfall at his inn ;

how next day, leaving Buckran and the groom snowed up at
Dalnacardoch, and taking only a small supply of raiment
with him, he and the other passengers toiled from breakfast-
time to sunset through that most formidable of the Highland
passes—the defile of Drumouchter ; how one of the
pedestrians, a sturdy sheep-farmer, would sometimes come
to the help of two young school-girls who were of the party,
lifting one in each arm through the heaviest drifts as if they
had been a couple of sheep ; how, after reaching and resting
at Dalwhinnie, they made their way finally to Inverness on
a snow-carriage ; and how Buckran and the dog-cart did
not turn up for nearly a fortnight after.

Inverness now became for a short while his headquarters.
There, as he writes, he had " long proses " with the Provost
of the town, who was factor for the Tarradale estate. He
went over the property, and "tasted" its soil with worthy
Provost Brown of Elgin, who pronounced it to be " good and
sharp." Like other Highland estates of the day, the land
was miserably farmed. We can picture the young laird,
mounted on Buckran, and riding among the wretched hovels
of his crofters. Little about the place itself, save that it
was his own birthplace and his father's choice, offered any
opposition to the design he had half-formed of selling the
estate. In his journal the following passage occurs : " When
the whole of the poor little tenants came round me and said
they would willingly pay any rent which their interpreter
into English, Rory M'Lennan, said ' so just a man as
the Provost would award,' I could not find it in my heart
to turn them adrift, though I knew them to be wretchedly
bad farmers, who hitherto had only paid their rents through
illicit distillation of whisky." Whether it was prompted by

mere good-nature or by youthful impatience, this hasty
letting of the estate in the old way to poor crofters proved
in the end to be as bad a piece of policy as the young laird's
uncle, General Mackenzie, declared it to be when he heard
what had been done.   In a few years the rents got more and
more into arrears, until the estate was gladly sold off.

Near to Tarradale lay the lands of Ferrintosh, the property
of Forbes of Culloden, to whom and his heirs, in consideration of services rendered and losses sustained at the time of
the Revolution, had been granted the perpetual right of
making and selling whisky at Ferrintosh, duty free.   The
temptation offered by such a traffic was too great to be resisted
by the tenantry of the other estates in the neighbourhood,
who readily found a sale in Ferrintosh for the whisky they
had privately distilled in their cabins or in lonely hollows of
the moors.   As a consequence of such extensive evasions of
the Customs, it became at last necessary to abolish the
privileges granted to Ferrintosh, the sum of £21,500 being
voted by Parliament in 1784 by way of compensation.   But
no Act of Parliament could readily change habits which
entered so largely into the life of the peasantry of that far
Ross-shire region.   And so the young laird of Tarradale had
to wink at the distillation, and pocket his rents, or at least
such proportion of them as he could secure.

Two Parliamentary elections occurred while he was at
Inverness, one of them for his own county of Ross-shire,
in which he took part on the side of the Tory candidate.   He
notes that at one of the election dinners he had the old chief
of Glengarry opposite to him.   "I saw," he writes, " that he
several times fixed me with his fierce grey eyes and bushy
eyebrows, and when the dinner was a little advanced, he put

his hand across the table, and leaning over said loudly to me, ' Ye 're welcome, sir, to the land of your fathers ; may you never desert nor forget it,' giving me a Highland grip I can never forget."

We may believe that a relative of Donald Murchison would not fail to receive a hearty welcome. Most of his time, indeed, during this visit to his native district, seems to have been passed in the enjoyment of the hospitalities of his friends and acquaintances—fishing, shooting, and hunting, and abundant festivity.

While amid such desultory employments and amusements time had been creeping onward with Murchison in Ireland, in London or elsewhere in England, and now in Scotland, events of world-wide importance had been shaping themselves in the Peninsula. Step by step Wellington had driven the French armies out of that part of Europe ; Napoleon's prestige had fallen, and at last came his abdication and retreat to Elba. Our young military aspirant says of himself that he was " for ever bewailing his fate at not being at his real work in the Peninsula." The campaign, however, had ended without his ever having had a call into active service, and now on the peace of 1814 he saw the final blow to all his hopes of military fame. As his uncle threw up his Staff appointment, he himself became a captain of the 36th on half-pay, his battalion having been promptly reduced. London became again his headquarters.

Of this part of his life the following notice occurs in his journal :—" In 1814 I was in London, living gaily at Long's hotel with a set of young dandies, dining now and then with Alexander Woodford of the Guards, at St. James's Palace,

when the announcement of the arrival of the foreign
Sovereigns (Russia and Prussia) set all the metropolis in a
ferment.   I galloped out with many others to Shooter's Hill
to see the Emperor Alexander in his little droschke, with
his bearded Russ on the box, and certes, though there was
no state reception, he was heartily cheered, escorted by a
joyous cavalcade of well-mounted English gentlemen.

"It being announced that the Regent would visit the
Opera accompanied by his imperial and royal guests, every
cranny was bespoke, and I got a good central post in the
pit; for in those days there were no stalls (and no shopboys
and tradesmen ever went to the pit then).   The reception of
their Majesties was of course most enthusiastic.   They were
really welcomed as our liberators from Gallic tyranny.

"Suddenly there arose a sort of semi-applause, followed
by murmurs, with some disturbance.   It was the Princess of
Wales, who had just entered a box directly facing that of
the Regent, and, as if she came to defy him and try her own
strength, she came forward in her hat and feathers to show
herself.   A few cries were got up for her, amidst loud murmuring at this unseemly attempt to disturb unanimity on
such an occasion.

"Then it was that the Regent, on whose countenance I
had my eye fixed, rose, and taking the Emperor and King
on his right and left hands, advanced gracefully to the front
of the royal box, the three personages bowing three times to
the audience.   The appeal was electric: the roar of applause
lasted for minutes, and the Princess was so discomfited that
she no more showed in the front of her box during the
evening, and retired soon to her *petit souper* and her
clique."

In the crowd of English travellers who eagerly availed themselves of the reopening of the Continent, Murchison found his way to Paris in the beginning of November 1814. He remained there for some weeks, which he employed with the most laudable assiduity in trying to make himself as French as he could.   He dined and spent much of his time in a *pension* where no English was spoken, took lessons in dancing from one of the leading teachers, frequented the theatres, passed many an hour over the pictures in the Louvre (for he was now beginning to aspire to be a connoisseur in art), was presented at Court, and in company with his old friend and fellow-aide-de-camp De Clermont, who had returned to Paris with the Restoration, saw everybody and everything which had any interest for " a young man about town."   There occur among his memoranda notices of the actors and the acting at some of the theatres.   " I could not," he says, " quite get over the solemnity and monotony of the French rhythm at the Théâtre Français, where I went, book in hand, to hear Talma in Corneille's ' Cinna,' supported, as he was, by Madlle. de Rancour and by Georges.   It was gratifying, however, to see how he first broke the sing-song by his imitation of Kemble and the English style by ejaculations and stops in the middle of some of the long lines of Racine.

" The best actor of high comedy I ever saw was Fleury. Having been taught before the Revolution, he was every inch a gentleman, and his countrymen of good taste said despondingly of him, ' C'est le dernier des Français qui sait porter l'épée.'   When I saw how vulgarly most of the other actors of the revolutionary breed dressed and acted, carrying their swords like butchers' knives, I felt the truth of the aphorism."

His mother was then living at Tours, and Murchison paid her a visit there. His chief companion there seems to have been Francis Hare (elder brother of Augustus and Julius), whose versatility and dash captivated him, and with whom he made excursions. Among other places, they visited together Poitiers, where Hare introduced him to Walter Savage Landor, then resident at that place. " Landor lived at the summit of a large central tower, which overlooked the whole city, and there we found the impetuous but warm-hearted philosopher ensconced in a library filled with all the most curious old French works, Rabelais being his special favourite. He and Hare held a disputation on Louis the Eleventh and his doings, as we looked down upon the remnants of the palace of that craftiest of all the French kings."

In such pursuits the last weeks of 1814 and the first two months of the following year passed away, until at the beginning of March he found himself again in Paris on his homeward journey. The morning after his arrival, his Swiss servant roused him with the momentous tidings, " Napoleon has landed in France !" The following narrative of this part of his experience is given by himself :—" To jump up, hurry on my clothes, rush out to the Café, already full of anxious and inquiring faces, was my first movement ; then to read the morning papers, most of them trying to make light of the affair, and saying it would be all soon put down. Next came reports that he had capitulated ; then that he was advancing to Grenoble. Right and left the English now were eyed inimically in the streets, low and vulgar officers elbowed you, and things became mightily unpleasant in the course of that day. On the following day, when more news had arrived, hopes were up,—the garrison at Grenoble had re-

sisted, and Napoleon's cause was lost; then a camp was to
be formed at Melun, and the Duc de Berri was to command
it; the Maréchal Ney having sworn fidelity to Louis XVIII.
This last, which was true, seemed the best chance, for Ney
was beloved by the soldiers. Then followed a review of all
the royal guards and regiments in Paris, 10,000 or 12,000
men, in the Carrousel in front of the balcony of the Tuileries,
in which the fat old Louis waddled out in his velvet boots to
be saluted by the loyal troops.

"I attended on that occasion, and never saw such a
farce. The soldiers of the line surrounding the National
Guards were all cracking jokes with each other; and though
they still wore the white cockade, they were evidently all
dying to mount the tricolor."

He went to see his friend at Court, the young Comte de
Clermont, and found him fully aware of the fact that the
army would not stand by the King, and that resistance was
therefore hopeless. Evidently Paris was no longer a desirable domicile for an English officer. De Clermont advised
him to leave at once. The English visitors were already in
rapid flight thronging the usual road to Calais, and hiring
every available conveyance that would take them to the
coast. Captain Murchison rightly conjectured that by making a detour by way of Béthune and St. Omer, he would
have some chance of securing post-horses, and reaching
Calais. Not without some risk, however, could English
travellers make their way along the roads of France at that
time. Coming out of Béthune he met the head of an infantry regiment, which, from the narrowness of the roadway,
had to pass the carriage in single file. " 'Que sont ces
Messieurs,' they cried out; 'Ce sont des —— d'Anglais.

Allons, renversez les à la baionette.' Drunken as they
were, and all in the greatest excitement, they had raised the
wheels, and were actually about to trundle us over into the
ditch of the fortress, and were unharnessing the horses, just
as the adjutant rode up and applied a thick cane to their
shoulders, and rescued us. We afterwards met with others
of these soldiers in detached parties, and in complete disorder, but we kept close shut up in our machine. At Arras
the captain of the guard sulkily let us pass the gates after
looking at our passports, saying, ' Et bien, je n'ai *pas encore*
reçu des ordres.' "

The war-clouds having once more spread over Europe,
there seemed now again some hope of obtaining active military service, and gaining coveted promotion. So the half-
pay captain of infantry determined at once to enter one of
the cavalry regiments which were to take part in the impending Belgian campaign. In doing so, however, he acted
without the advice and indeed against the wishes of his
uncle, General Mackenzie, who, vexed at this want of confidence, wrote to his mother that he considered the entering
into the cavalry as a " measure full of the most stupid folly."
Notwithstanding this protest, the exchange was made. Murchison joined the Enniskillen Dragoons, and seems now to
have looked forward with tolerable confidence to a chance
of distinguishing himself. But even though he had the
promise of employment from the Colonel, who was his personal friend, he was once more fated to disappointment, and
the predictions of his uncle proved too true. Six troops only
were ordered out, and every one of the service captains insisting on going ; he had no alternative but to equip himself
with uniform and horses, and repair to the depot at Ipswich,

Events crowded rapidly upon each other during the
hundred days,—Ligny, Quatre Bras, and, lastly, Waterloo.
Then fell Murchison's hopes of an active military career.
The war was at an end.   Europe had now been so worn out
with fighting that no new campaign was likely to take
shape for many a long year to come; and, in the meanwhile,
he had no brighter prospect than the *ennui* of half-pay.

He was now, however, nearing the event which, in the
end, proved the turning-point of his career.   His mother,
like other English residents in France, had deemed it prudent to quit that country after Napoleon's return, and had
settled for a little at Ryde, in the Isle of Wight.   Thither
her son went to visit her, and there, through the introduction of Miss Maria Porter, he made the acquaintance of
General and Mrs. Hugonin of Nursted House, Hampshire,
and their daughter Charlotte.   This young lady was, to use
his own words, " attractive, piquante, clever, highly edu-
cated, and about three years my senior."   He first met her
early in the summer of 1815, and, on the 29th of the
following August, in the romantic little church of Buriton,
in Hampshire, they were married.

Want of success in the military life had disposed Captain Murchison to look on that career with less enthusiastic
feelings than those of earlier years.   He had even gone so
far as to think of retiring from the army; and now this
half-formed intention received a stimulus from two sources.
His wife, herself the daughter of a soldier, had experienced
some of the discomforts of a soldier's life, and discerning in
her husband qualities of a higher kind than would be likely
to be called out by the routine of barrack-duty, seconded
his own inclinations.   But perhaps the more immediate

cause of his final determination was an order to join his regiment at Romford barracks. To take his bride there, that she might share the dulness with which his experience at Horsham and Armagh had made him only too familiar, was a most distasteful prospect; so at last he made up his mind and sent in his resignation. His commanding officer remonstrated with him, but in vain. He stuck to his purpose. After eight years' service he finally retired from the army and gave up all those visions of military glory which filled his whole soul in the old Marlow days.

It is evident that, up to this period of his life, Murchison had not in any way given promise of future distinction. He would have been noted as merely one of the gentlemanly, intelligent, but by no means brilliant young officers, so plentiful in the British army. To one who judged him merely by externals, he would undoubtedly have seemed little else than a military fop, and he used in later years to confess that such an estimate would have been tolerably true. The circumstances which were to call out his special qualities of excellence had not yet arisen. Full of health and bodily activity, he had from the beginning looked on the military profession rather as an outlet for that part of his nature than as a career requiring any special mental training. In those days, indeed, professional study was not much in fashion in the army. After quitting Marlow he does not appear to have given himself in any degree to acquiring further knowledge of the principles of the art of war. In his journals there can be found no trace of professional study, nor indeed of solid reading of any kind. His leisure, which must often have hung heavily on his hands, was spent, as we have seen, in active field-sports, in feats of bodily exercise, or in gratifying that love of display which led him into culpable extravagance; so that when he quitted the army, there was little to look back upon with unmingled satisfaction in that introductory part of his career. He had entered the service with high hopes of distinction, but by a series of unfortunate circumstances, and through no fault of his own, he had been grievously disappointed. The war had now come to an end, and with it went his visions of rising to distinction in a campaign. He had not qualified himself for distinction in any other way, and we can well imagine how he should have turned aside at last almost with repugnance from a career which at the beginning seemed to promise all that he most desired.

Hitherto he had lived at his own free will. From this time he came under the influence of a thoughtful, cultivated, and affectionate woman. Quietly and imperceptibly that influence grew, leading him with true womanly tact into a sphere of exertion where his uncommon powers might find full scope. To his wife he owed his fame, as he never failed gratefully to record, but years had to pass before her guidance had accomplished what she had set before her as her aim.

The wedding over, Murchison took his bride north to show her the Scottish Highlands, and to visit his friends and relatives there. Of course he did not fail to lead her over the paternal acres of Tarradale, and show her some of the scenes where his ancestors had distinguished themselves. Among other houses they visited that of an old lady, a grand-aunt of his, who had intended leaving her estate to him or his brother Kenneth, but unfortunately for him, as she confided to his young wife, " he had too much of the Baillies about him," his grandmother having been a Baillie;

and so the estate, which would have been a welcome addition to the badly paid rents of Tarradale, passed into other hands. Late in October, and in a storm of snow, they migrated southwards again.

Having given up one fixed employment the retired captain of dragoons began to look about for another. It will hardly be believed by those who only knew him in his later years that he now seriously thought of becoming a clergyman. In this proposal, as in his choice of a military profession, it seems to have been mainly his love of bodily activity and open-air exercise which swayed him. He says of himself,—" I saw that my wife had been brought up to look after the poor, was a good botanist, enjoyed a garden and liked tranquillity; and as parsons then enjoyed a little hunting, shooting, and fishing without being railed at, I thought that I might slide into that sort of comfortable domestic life." Among the letters which he preserved there occurs one from a friend whom he had asked to make inquiries for him, and who went into the question in the most earnest and business-like manner. This correspondent urges the necessity of getting a Greek Lexicon, and suggests the name of a clergyman who might be of service in helping the aspirant for holy orders to read the Greek Testament. So earnest is he about the Lexicon and other heavy tomes, that he insists upon Murchison's having them conveyed separately if he could find no room for them in the carriage with which he proposed to make a journey to Switzerland.[1]

[1] The gravity with which the question was viewed may be gathered from one or two sentences taken from this letter :—" In consequence of the peace we may expect an irruption of officers into the Church, which may produce an additional strictness of regulation. I am not aware in what time a degree may be taken at Cambridge; any Cambridge man

Fortunately for himself and his possible parishioners this notion soon died away. But while still undecided about entering the Church he resolved in the meantime to see a little of the world with his wife. The winter was accordingly passed at Nursted House, in diligent preparation for a long and leisurely tour on the Continent. He had already attained considerable proficiency in French. As the tour was to be extended into Italy, he now set diligently to work to acquire further knowledge of Italian, and to read a quantity of literature treating of the scenery and history of Italy. Probably this was the most industrious winter he had yet spent; for he had now a definite incentive to work, besides the example and co-operation of his wife. A day now and then with the Hambledon fox-hounds, or old Tom Barham's beagles at Petersfield, or with his gun and his father-in-law at home, kept him from suffering from such an unwonted application to books.

would tell you. The examination is almost nothing. Not so at Oxford, where the whole system would present to you considerable difficulty." " Surely as you are so well known in Ireland you might find a favourable bishop in that country, and the journey would be the work of a fortnight. At any rate, pray do not give up your excellent plans, *dégoûté*." " I will in your absence, without mentioning your name, make every inquiry I can. The stability and well-being of our Church depends so much upon the respectability and fitness of its ministers that we can only quarrel with those forms and preliminaries to ordination when they come in competition with our own favourite wishes " !

In a note-book of 1815 there occurs a most formidable list of books which it seems Murchison had jotted down with the intention of using them in his proposed clerical education. They are in Greek, Latin, French, Italian, and English, and with his characteristic methodical habits he has classified them under various heads, as " Religion," " Eloquence," " History," " Belles-Lettres," etc. etc.

CHAPTER V.

ITALY AND ART.

WITH the proposal of a country parson's lot still undecided, and indeed with no settled plans for the future, Mr. and Mrs. Murchison had determined in the meantime to spend a year or two abroad. This resolution had been, in some measure, forced upon them by the state of their finances. The Tarradale rents, never very well paid, even at the best, had almost ceased to yield any income, and times were so bad that the tenantry petitioned for alleviation. His revenue from other sources was not great, certainly not enough to enable the young laird and his wife to live comfortably in England. It was sufficient, however, to permit them to enjoy comfort, and even elegance, in Italy. So that, until some decision had been come to regarding the fate of the Highland property, a sojourn on the Continent was deemed absolutely necessary.

This enforced exile, however, proved in the end eminently advantageous in other than a pecuniary sense. Mrs. Murchison had shrewdly discerned her husband's true nature and the way in which it should be developed. She saw that with his tastes and habits he would be far less likely to break off from a useless kind of life at home than if placed amidst a totally new set of pursuits and acquaintances abroad.   And thus the continental sojourn was planned and the notes of travel were prepared that the foreign scenery and associations should act as powerfully as possible on his mind.   It was a sagacious experiment, and it succeeded.   In this chapter we have to trace how it was carried out.   Its fruits will appear in later pages.

On Good Friday 1816 the young pair sailed from Dover, and taking with them their own carriage, posted by easy stages from Calais to Paris.   About a year had elapsed since the hurried flight from that capital noticed in the preceding chapter, and now the masons were found to be busy on scaffolds removing the letter N from the public buildings.   On that previous visit Murchison had made himself tolerably familiar with the contents of the Louvre, then enriched with the spoils of Europe ; and his first object now was to see how the galleries looked after having been made to yield back their treasures to the rightful owners.   He was " astonished to observe how rapidly the vacant places had been filled up, and not unfrequently by good old Italian pictures, which had also been *stolen*, but which not having been exposed in the Great Gallery were not known to exist in France."

During a most systematic tour of the sights of Paris he attended a meeting of the Academy (which many years later was to enrol him among its foreign members), and saw Cuvier for the first time, who declaimed upon the influence of the sciences on the common occupations of man, and upon the leading share which France had taken in promoting this influence—a share which would have been yet greater had it not been thwarted by the *perfide politique* of England.

From Paris they journeyed in the same leisurely way by Dijon to Geneva. Though Murchison had as yet shown no special interest in science, he now began to make the acquaintance of scientific men in the places he visited, and paid some attention to their museums. At Geneva, for example, he met among others Pictet the naturalist, and De Candolle the botanist. He found too that "the same rigid solemnity was observed there on the streets on Sunday as in Edinburgh—all demure and starch." "I induced," he writes, "good Madame Peschier to go a drive (and we had been at morning service), but when descending the steep street from the house a grave-looking churchwarden, who was going to afternoon service in his black silk stockings and a gold chain, came up to us, and holding out his watch, pulled up our horse, and exclaimed, 'Madame Peschier, je suis étonné! vous auriez dû connaître que pendant les heures de l'église on ne va pas en voiture.'"

The summer was spent at Vevay, where he took a little villa. His wife's ancestors had come into England from that part of the Pays de Vaud about a hundred years before. She found some distant relations there who made the sojourn at Vevay a memorably pleasant one. Many excursions were made to surrounding parts of Switzerland, the ladies usually driving or riding, while Murchison himself delighted in keeping pace with them on foot. Leaving his wife in charge of her Swiss cousins, he undertook some feats of pedestrianism of which he used to boast in his old age. On one occasion he walked 452 miles in fourteen days, on

the last day of which excursion he accomplished 57 miles.
In another excursion to Mont Blanc he walked 120 miles in
three days.   Such rapid marching is suggestive rather of
exultation in bodily activity than of intelligent appreciation
of scenery. Yet his singular power of rapidly seizing the main
features of a landscape enabled him to carry away some vivid
impressions of what he saw, and even to note some of the
details.   In his itinerary journal, he speaks of the Grindel-
wald glacier as a " river of ice," and among his notes there
occurs a detailed narrative of the processes in use at one of
the Swiss salt-mines.

An interesting episode of their life at Vevay may be
noticed here.   A terrific thunderstorm broke one night (13th
June) over the lake in front of them, and, roused from sleep,
they sat watching from the window a scene never to be
forgotten.   Some months afterwards they read at Rome the
now well-known lines in the then newly published Third
Canto of *Childe Harold* :—

> " And this is in the night !—Most glorious is the night,
>   Thou wert not sent for slumber ! let me be
>   A sharer in thy fierce and far delight,—
>   A portion of the tempest and of thee !
>   How the lit lake shines, a phosphoric sea,
>   And the big rain comes dancing to the earth !
>   And now again 'tis black,—and now, the glee
>   Of the loud hills shakes with its mountain-mirth
> As if it did rejoice o'er a young earthquake's birth."

The passage recalled their experience at Vevay, and brought
to their recollection that they had met Byron walking from
Vevay to Clarens on the day before the thunderstorm which
he has immortalized.

The winter of 1816-17 was passed at Genoa, studying
Italian, and kindling a passion for art and art-galleries, which

a few months later was to burst into a most portentous blaze at Rome.   Murchison found opportunity too of practising his favourite exercise—walking, in which, as his notes record, he outstripped two young officers since known as intrepid travellers—Irby and Mangles.   In one of his excursions marine shells were noted upon some of the hill-tops, and he infers that these high grounds were once under the sea.

By the 21st of March, ere Holy Week began, the two travellers had reached Rome.   Owing to the cessation of the war and the reopening of the Continent, the city happened to be at this time crowded with strangers.

Established, however, in a private lodging in the Via Condotti, Murchison avoided gaiety, and became now a confirmed dilettante.   Day by day, accompanied and incited by his wife, he visited gallery after gallery, and church after church, making elaborate notes on the pictures and other works of art.   He seems to have left little in Rome unseen, and his jottings, written at a time when the profuse modern literature of "Guide-books" and "Hand-books" had not yet made its appearance, show a creditable degree of zeal and intelligence.   The general style and tenor of those art-notes and criticisms may be judged of from the following specimen of his journal :—

"*Rome, June 13th, 1817.—Palazzo Colonna.*—Four superb landscapes of Salvator Rosa (doubtful) ; marine views, with armed men and fishermen in the foreground.   The light and distances have the light of Claude, the foreground less of the savageness of Salvator than usual.   Two fine heads of Carlo Dolci, one St. Catherine, the other a saint chained.   Some good heads of Guercino, and a fine small piece or two by Conca.   Many good landscapes of Poussin in tempera, and one beautiful bluish landscape of Lucatelli, marine, with great depth : this is in his best style.   The Bella Cenci needs no description.   Guido is more expressive here than in his fine exuberant Madonna above stairs. There are two little Claudes, and a Titian, etc.   There are a good many pictures of the inferior and later Roman artists; some of these are pleasing.   Gaetano Lapis (1776), a scholar of Conca ; same light colouring, but no confidence in himself.   His best picture here appeared to me a Lazarus with Christ (doubtful).   The frescos of Stefano Pozzi in first room are bright and pretty (Turk smoking).   The column of Bellona (twisted) of *rosso antico,* with Pallas on the top, very beautiful. A Dead Christ by Franc° Trevisani (d. 1746. Sc. Rom.), not Angelo Trevisani (Venet. Sc. same epoch).   In this Christ the foreshortening is remarkable, the colouring Guidesco.   He was a universal imitator."

Of the acquaintances whom Murchison made at Rome the most notable was the sculptor Canova, with whom he had frequent intercourse at the house of Cavaliere Tambroni, then a sort of chief of art.   From his journal and a pencil note written late in life the following reminiscences of the sculptor are given :—

" When asked what he thought the most wonderful structure in Britain (for he had recently visited England), he at once replied, ' Waterloo Bridge.'   Of the antiquities in the British Museum he gave unquestionable precedence to the Ilissus of the Parthenon, preferring it on account of the inimitable schiena to the Theseus.

" He narrated to me how he overcame Buonaparte's obstinacy, who at first insisted that the great sculptor should represent him in marble in the garb of the conquering French General with cocked hat, straight cut coat, and top-boots—hunting-boots 'à l'Anglais.' Canova stood firm in refusing, and when he said to the future Napoleon, 'Then your Excellency must find other artists, and I can recommend both a tailor and a bootmaker in the Corso,' the Corsican at once saw a man of taste and genius must have his own way, and Napoleon came out in classical toga, etc.

" Canova was a very active man, and when debarred of his exercise by too much work in the studio, he was in the habit of jumping backwards and forwards over his modest bed, and, proud of his agility, he did it before me.

" This eminent sculptor passes an hour or two every evening at Madame Tambroni's ; at nine o'clock he invariably retires. Had a long conversation with him the other night. He observed to me, that when in London nothing offended his eye more than the smoky brick houses with clear painted windows, and was surprised they were not all white-washed. He spoke of the absolute necessity of our having a museum superior to that of Somerset House. The education of English women delighted him, and he the more regretted the state of his own compatriotes. He asked why all the English began their Italian with Dante and Boccaccio. Metastasio seems to be his favourite author. The style of the one in literature is similar to that of the other in sculpture—both chaste, classical, graceful, and full of pathos. He said of Metastasio's critics, ' Quei che lo criticano, lo leggono ; e poi piangono.'

" In Canova's studio no one appears more conspicuously than the distorted Giaccomino. Ask him where he has been, and he answers, ' We have been modelling above stairs, il cavaliere ed io.' Giaccomino was a poor, good-humoured countryman, whom Canova employed as a sort of lower servant in the workshop. He sometimes hands the morsels of clay to his master whilst he is forming the cast, and from hence Giaccomino concludes that at least half the merit is his own. He freely canvasses every new attitude, and Canova says, ' E mio maestro Giaccomino,' and always asks for his opinion upon any new work. In these little traits the playful *bonhomie* of the great sculptor is pleasingly exhibited.

" To judge of Canova's simplicity, examine his house. You will find every article neat and appropriate; no luxury, but the utmost cleanliness and regularity—doubly delightful in so filthy a country. Two of his bedrooms are ornamented with his own paintings. During the French invasion he occupied himself for eighteen months with the brush and palette. The compositions are in general just what you might look for from the graceful mind of the artist—a sleeping Venus intruded upon by a peeping Satyr, Venus with Cupids, etc. The colouring is Titianesco, and very wonderful. These pictures have already the mellowed tone of the colouring of the old masters; and a head of an old carter (a portrait from life) is painted expressly to deceive as an antique.

" Madame T. related to me, that when Canova first imagined his group of the Graces, he happened to be in the country visiting the Cavaliere T. Here there were no fine models, but females must be found. Accordingly, two large and fat female domestics of Madame T. were paraded, who, with herself, formed the graceful trio. Their attitudes must have been most diverting to Canova whilst he drilled and practised them. Canova is now nearly sixty years of age, yet constitution and physical powers are such that he can jump over his bedstead à pie pari, and can extend a prodigious weight with his arm."

Three months specially given up to fine art soon passed away in Rome. The journal in which the record of that time was so elaborately chronicled is, however, more a dry inventory of what the writer saw than of what he thought and felt.[1] Now and then he varied his researches by an excursion into the country, but an unfortunate event cut short these occupations. His wife caught a malaria fever, and became so ill that he despaired of her life. Rallying at last, she was able to be moved from Rome at the end of June to seek a change of air and the sea-breeze at Naples.

Full of details though the journal is regarding the stay at Naples, little occurs of any general interest, or which throws any fresh light upon Murchison's own character and development. He visited, of course, all the usual places of resort in that neighbourhood. The nearer excursions were made with his wife, but in company with a military friend he accomplished a series of boating expeditions to Pæstum, Capri, Ischia, and Procida, seeing a good deal both of scenery and of Italian life outside of the ordinary beaten track of tourists. He was lucky enough to come in for an eruption of Vesuvius, and ascended the

[1] No mention occurs in the journal of his havin at this time made the acquaintance of Mrs. Somerville and her husband. In her charming *Personal Recollections* (p. 122), she thus alludes to the incident:—" Our great geologist, Sir Roderick Murchison, with his wife, were among the English residents at Rome. At that time he hardly knew one stone from another. He had been an officer in the Dragoons, an excellent horseman, and a keen fox-hunter. Lady Murchison,—an amiable and accomplished woman, with solid acquirements, which few ladies at that time possessed. . . . It was then that a friendship began between them and us, which will only end with life."

mountain when a current of lava was streaming down its
side.   To get the better view he made the ascent by night,
and there being no moon, had an impressive view of the
huge lurid crater, with its rocket-like showers of red-hot
stones, and scrambled over the hardened but still hot surface
of lava to see where the molten mass came out in a glowing
stream from the side of the cone.   His notes of this visit
are simply those of an intelligent and interested spectator;
they betray not the slightest geological predilection.

In Naples, as in Rome, his favourite occupation was to
visit the art-galleries and altar-pieces in the churches, and
to write out detailed descriptions of the pictures and statues
in his journal.   Even the sight of the miracle of the
liquefying of the blood of St. Januarius could hardly inter-
rupt the art-fever; for though the saint gratified the curiosity of the two travellers and the prayers of the orthodox by
thawing the blood in three minutes instead of keeping them
waiting for hours, the enthusiastic but irreverent dilettante
writes in his diary, " We slipped away from the altar to
admire, not the works of the saint, but the sublime repre-
sentations of them by Domenichino."

Early in October 1817 Murchison returned with his
wife to Rome, and wintered there.   Art again became his
absorbing pursuit.   Every gallery was once more visited,
fresh notes were duly entered in his journals.   His criticisms,
after a few months of experience, are spiced with the dog-
matism and the pet phrases of a confirmed connoisseur of
many years' standing.

Having taken his fill of art and the galleries, Murchison
next set to work with equal industry upon the antiquities
of Rome.   A good part of the winter of 1817-18 was spent

in sedulously tracing the lines of the several walls, and the position and remains of temples and public buildings. He entered with his characteristic zeal into the disputed localities of the Forum, and not content with reading such of the lucubrations on this subject as he could reach, he wrote in his journal voluminous comments of his own upon previous writers, and gave the observations he himself had made, with the conclusions to which they had led him. He revived his long disused and never very familiar Horace, Virgil, and Juvenal, with whose allusions to Rome and Roman sites he interspersed his notes. The following extracts may suffice as a specimen of the style of these antiquarian memoranda :—

" *Grotto of Ægeria.*—

' In vallem Ægeriæ descendimus atque speluncas
Dissimiles veris.'

In Juvenal's day great had been the alteration of the little consecrated grot of old Numa, which was of tufa. Now this is the only tufa cavern in this valley. In the time of Cicero the simple old cavern was decorated with marbles and statues, and became ' dissimiles veris ;' now the present work as extant, and the reticulated brick, are all of the latter end of the Republic. The recumbent statue of the man proves nothing, as the figure evidently represents a river (viz. the Almo, which rises here), from the urn under his arm. The goddess might have been placed in the same niche above him. Everything marks this distinctly to have been the sacred spot.

" *Templum Rediculi.*—Positively a temple and no tomb, Mr. Eustace.[1] The cella and component parts remain.

---

[1] He refers to Eustace's *Classical Tour*—a work which he studied

Hannibal might first have appeared here, and then making a detour might have encamped on the other side of the town. It has been rebuilt in the age of Severus. Four styles of architecture are to be observed in it.

"*Baths of Caracalla.*—Double purpose, bathing and amusement. The baths were below ground, and had no communication with the halls above, no staircase having ever been discovered. The great portico to the west, with the various little chambers, was a quarter for troops, from which a spiral staircase conducted to a terrace above for parade and exercise; but no communication took place by doors between these chambers. The grand central mass of building was entirely enveloped and shut in from sight by a still more vast pile. These covers or cases for buildings were common to the Romans, for in this exterior an uniform height was preserved, which hid all the inequalities of height and construction of the internal pile. This will account for the arches of different elevations. . . .

"*Cecilia Metella.*—Republican work : crowned with an entablature, and formerly with an attic and a dome.

"*Forum Romanum.*—

> ' Vespertinumque pererro
> Saepe Forum.'   Hor. *Sat.* i. vi.

Old Horace could not have enjoyed his evening walk there more than I do, and one great delight consists in the imagining that I behold some relics of those very buildings which he admired. Away then, ye cold sceptics who drive everything to such an extreme that at last ye begin to doubt whether ancient Rome did really exist here, or

before leaving England, and which he seems to have carried about with him in Italy, and to have found as unsatisfactory a guide as Byron did. (See Note xxxii. to Canto iv. of *Childe Harold.*)

whether the Tiber may not have changed its course! They will tell you (even Nardin and others) that most part of the columns have been re-erected in subsequent ages on or near the spot where they had fallen or been pulled down. But, oh ye learned sceptics! what Pope, Antipope, or Goth, may I humbly crave, would ever have had the genius of architecture and the love of classical remains impressed so deeply on his mind that he should wish to raise up broken entablatures of colossal size, and mutilated columns, in order that he might be called a man of taste? If, therefore, none of these re-erections took place in the dark ages, which I think any reasonable man will allow, we can have little difficulty in proving that such attempts have not been made since the revival of letters in the fifteenth century. Private and public history are both silent on this point, whilst on a number of trivial little subjects, such as that Lorenzo di Medici robbed the Dacian captives on the Arch of Constantine of their heads, and other similar facts, we have abundant details."

While this antiquarian fever lasted, he made an excursion on foot to Præneste, walked along ancient highways now deserted, but still level and unbroken, looked into the memorable crater-hollow of the lake of Regillus, with a half-antiquarian, half-military, but in nowise geological eye, remarking that the allies had much the better position, since the Romans had to charge up hill; scrambled up to the Cyclopean walls of Præneste, and from the summit of the town let his eye wander over that marvellous landscape, so rich in association, from the far southern Apennines away across the Alban and Volscian hills, into the limitless Campagna.

About the middle of March (1818) Mr. and Mrs. Murchison quitted Rome for a leisurely journey homewards. At Florence they lingered for three weeks, chiefly among the galleries and museums. Again his note-books teem with descriptions and criticisms of the pictures, his later studies at Rome having given him greater confidence than ever in his judgments on art. Michael Angelo receives a special measure of his critical wrath. More interesting is it to mark that among his notes of Florence some space is given to an account of the Museum of Natural History, particularly that portion in which the successive stages in the growth of animals were illustrated. From Florence the journey led by short stages, and with many a halt, to Bologna, Modena, Parma, and Turin, thence by Mont Cenis into Switzerland, and then by way of Lyons to Paris, and so home.

Rather more than two years had thus glided away on the Continent; two memorable years in Murchison's life. They taught him, in a way which would have been little likely to occur to him at home, the superiority of such pursuits as called for the exercise of thought and taste over the more frivolous employments of barrack-life. It is true that his wife was always at his side to share in his pleasures and incite him to further perseverance in the new line of occupation. But her influence was little needed after the first decided tendency had been given to his inclinations. He soon became a far more enthusiastic lover of art than she, and must no doubt have often tried her bodily strength to the utmost in his hunt through churches and galleries for Guidos and Raphaels, Caraccis and Domenichinos, in all the stages and styles of each painter. For the time, he was absorbed in art and Roman antiquities. It was the first taste he had yet had of the pleasures of continuous intellectual employment, and he threw himself into it with all the eagerness and enthusiasm of his nature.

He had a natural weakness for display, which in his military days, as we have seen, took shape in fashionable clothes, horses, and the other extravagances by which a young man in the army contrives to get rid of his money. In Italy no such temptation came in his way. For the time he was left to the influence of his wife and his own better nature, with the result of receiving a deeper and better impress on his character from these two years abroad than from his eight years in uniform. Unconsciously he was sowing seeds which would in after years bear fruit of a very different kind. Through art he first realized the advantage of a distinctly intellectual life over one of mere desultory gaiety. It was not art which was to furnish his future stimulus, and, as we shall find, it did not even suffice to keep him from relapsing into some of his old ways when the temptation came back again. But his art-studies in Italy formed the starting-point of a new life for him, and led the way to all the work and honours that were to come.

CHAPTER VI.

WHEN Murchison and his wife found themselves in England again, two questions pressed upon them for immediate solution : Where were they to take up house ? and, What were they to do ? In spite of Mrs. Murchison's fortune, money was not so plentiful with them as they wished. The Tarradale tenants, owing to more stringent prohibition of illicit distillation, found many excuses for evading the payment of their rents, so that although the young couple could live comfortably enough in Italy, there seemed some difficulty in the way of their setting up house at home in the style to which they had all along been used. The rent of the property was at this time a little more than £500, but probably not more than about the half of that sum could be collected. The long-threatened sale was therefore now finally resolved upon, and in August 1818, for £27,000, Baillie of Dochfour became the purchaser. Immediately after his return from abroad Murchison went north alone to make the concluding arrangements, and from that time ceased to be any longer a Highland laird.

Having thus got rid of the troublesome tenants in the north, he had next to find a home somewhere for his wife and himself.   Mrs. Murchison's grandfather, a veteran of the Flanders wars, had passed the last twenty years of his long life in an old mansion at Barnard Castle, in the county of Durham.   This house, now tenantless, was chosen, and there Murchison set up his first *ménage* in England.

The change from the pursuits and sights of Rome and Naples to the dulness of a little country town in the north of England could not but prove a sore trial to the lately developed tastes of the retired Captain.   The old General, whose house they now occupied, had been a favourite in the district, and for his sake at first, and afterwards for their own, the new-comers had a hospitable reception from the county-folk of the neighbourhood.   But receiving calls and paying them was hardly occupation enough for any reasonably active creature.   Art-studies were no longer possible; his wife's gathering of plants and minerals had not yet sufficed to show him what a scientific pursuit really was; there seemed but one path of escape from insufferable *ennui*, and Murchison chose it.   He took heart and soul to field-sports, and became one of the greatest fox-hunters in the north of England.

For five years this desultory life lasted.   It seemed as if the influence of the foreign tour had vanished, and left no sign.   At some of the houses of the neighbourhood—Rokeby, for instance—guests distinguished for culture and literary or scientific eminence used from time to time to be gathered, and in these gatherings Murchison and his wife gladly took part. They only just missed Sir Walter Scott.   They formed an intimacy with Sir Humphry Davy, and made the acquaintance of other notabilities.   These were pleasant inter-

ludes, and helped to vary a little the dulness of Barnard
Castle and the monotony of hunting. But field-sports continued to be the main business of life, since they furnished
the readiest outlet for that exuberant bodily activity which
had all along formed one of Murchison's special characteristics.

As a diversion from these more ordinary and engrossing
pursuits, he on one occasion of a contested election for the
county of Durham took an active part on the Tory side,
scouring the country far and wide on horseback for voters,
bringing them up to the poll; but in the end beating an in-
glorious retreat with the unpopular candidate, amid showers
of cabbages, rotten eggs, and other electioneering missiles.
A further variety was found in an occasional excursion to
Scotland, or in visits to sporting friends in the north of
England.

It was not without concern that Mrs. Murchison marked
this relapse into that purposeless kind of life from which her
husband seemed for a time in a fair way of being weaned.
She had some knowledge of botany, and had induced him in
the course of their walks and excursions to assist her in forming a herbarium. But she could not make him a botanist.
While residing in the north of England she took to the study
of mineralogy, and made some progress in collecting and
distinguishing some of the more common minerals found in
that part of the country. Her husband looked on and helped
her where he could; but neither was mineralogy the kind of
pursuit to enlist his sympathies, and call out his special
powers. "The noble science of fox-hunting," he says of
himself, "was then my dominant passion, and as I had
acquired a little reputation in the north as a hard rider, I

resolved to play the great game, increase my stud, and settle for a year or two at Melton Mowbray, in Leicestershire."

Instead of calming down, therefore, the hunting fever broke out with renewed virulence. The migration southwards duly took place, to the great mortification of his wife, who had reason to dread the effects of the change both upon his character and his purse. He rented a good house at Melton Mowbray, kept eight hunters, a horse for his wife, and a hack, and subscribed £50 a year to a pack of hounds. "These and other expenses were," he says, "more than enough for my means. Thus I was led to speculate by investing in foreign funds, and obtain an income of £2000 per annum, which, with occasional drafts upon my 'floating capital,' kept us going."

He paid a visit to the north of Scotland in 1822, and his arrival in Edinburgh happened to coincide with that of George IV., whose entrance he witnessed from the Calton Hill, noting especially the beaming face and white hair of Walter Scott as he marched jauntily along in front of the royal carriage.

Back at Melton, he recommenced the earnest business of the winter by resuming his place at the hunt, and indulging in further gaieties.[1] The following reminiscences of this time were written late in life :—" On Sundays, after six days' hard work, we were necessarily very sleepy, and on one occasion when the sermon was preached for the Missionary Society, and the parson went on to describe the life of the savages to be Christianized—hunting all the week, and lying

---

[1] By way of compromise, apparently, and in compliance with his wife's more literary tastes, he kept his elaborate daily hunting journal this winter (1822-3) in French.

exhausted and sleepy in their houses,—all the ladies' eyes were turned upon their drowsy mates."

" On one occasion I gave a dinner, and invited Scotchmen only, viz., Elcho, Graham (now Duke of Montrose), Grant, Melville, etc.; and as I could find no blacksmith to singe the head, I performed myself in my own stable-yard, to the great amusement of the groom and helpers."

" I was the only person who regularly smoked at the covert-side, or when they went away, and the fox was lost. On one of the latter occasions, and when Graham was casting and re-casting his hounds, and was unable to hit off the scent, he hollowed out sulkily, ' 'Tis no use trying to do anything when that —— pipe spoils the scent !'   So strong was the feeling then against smoking as a bad and ungentlemanlike habit, that when Fernley painted a picture which we, the subscribers to the pack, presented to Graham, I was at first represented on my brown horse Commodore, turning my head round, with a cigar in my mouth.   The cigar was afterwards, however, painted out.   The picture is at Norton Conyers, in Yorkshire."

Save gossip of this kind, with full notes of his almost daily hunts, and references to the companions with whom he rode, smoked, and dined, the visits which he and his wife occasionally paid, and the people whom they met on such occasions, no record of these five hunting years has been preserved.[1]   There seems, indeed, to have been little else to chronicle.   During the times of hard frost, when the usual

---

[1] One of his journals gives a detailed narrative of every hunt from 3d November 1821 to April 11, 1822, during which period he was 110 times with the hounds.   In his usual methodical style he has constructed a table with columns, in which is entered the work done by each of the twelve hunters which he used.

out-of-door occupations were interrupted, he would take once
more to books. On one of these occasions he seems to have
revived for a while his antiquarian tendencies by reading and
making extracts from Blunt's *Vestiges of Ancient Manners
and Customs in Italy and Sicily.* But the books were exchanged for the saddle when the weather suited again.

The letters written during these fox-hunting years to his
brother Kenneth, then in the East Indies, abound with grave
moral sentences on the duty of submission to our lot, and
the necessity for economy and care when our means are
small ! Yet they teem with tender affection, and show their
writer to have had an earnest love for his brother, with the
fullest interest in all that concerned him. The solicitude
with which he appears to have watched over a little niece
confided to his care and that of his wife, and the almost
fatherly delight with which he recounts all her ways and
her progress, betoken great tenderness of heart, with much
considerate feeling in the way of showing his kindness.

His wife had from the first truly perceived that at bottom there lay in Murchison something more than the character of a mere Nimrod. It was needful that his overflowing
animal spirits and bodily activity should find adequate outlet,
but she fully believed that when these parts of his nature
had in some measure spent themselves, the higher part of
his character would come to the surface. If he really had
any more intellectual tendencies than were required for fox-
hunting, he must needs in the end get tired of such unremitting application to that pursuit, and then those tendencies
would be sure to claim a hearing from him. And so it
came to pass.

Forty years after the time at which we are now arrived,

Murchison was sojourning for health's sake at the baths of Marienbad, in Bohemia, and penned there the following recollections of the events which brought his fox-hunting life to a close :—

" As time rolled on I got *blasé* and tired of all fox-hunting life.   In the summer following the hunting season of 1822-3, when revisiting my old friend Morritt of Rokeby, I fell in with Sir Humphry Davy, and experienced much gratification in his lively illustrations of great physical truths. As we shot partridges together in the morning, I perceived that a man might pursue philosophy without abandoning field-sports ; and Davy, seeing that I had already made observations on the Alps and Apennines, independently of my antiquarian rambles, encouraged me to come to London and *set to* at science by attending lectures on chemistry, etc.   As my wife naturally backed up this advice, and Sir Humphry said he would soon get me into the Royal Society, I was fairly and easily booked.

" Before I took the step of making myself a Cockney I sold my horses.   The two best were put up at auction in the ensuing autumn, after dinner, at the Old Club at Melton, and were brought into the room after a jolly dinner, Maxse acting as auctioneer.   In fact I threw them away, and Maker who bought the ' Commodore,' named him ' Potash,' as a quiz on me for taking so much of that alkali after our potations."

The decision to sell his hunters and renounce the expensive life at Melton was probably dictated more by a prudent regard to ways and means than by any special charms yet visible in the prospect of a life of scientific exertion.   At all events we find, that when the Melton establishment was broken up he did not immediately set up another, but went to reside for a time with his father-in-law.   The winter of 1823-4 was passed chiefly at Nursted House, and seems to have slipt away without much indication that he had resolved to change his main pursuits.   Were not the Hambledon hounds at hand, with old Parson Richards at their head, and Wyndham's drove pack careering in close column up the steep faces of the downs ?   Did not Up Park offer attractions in its pheasant covers such as few other pre-serves in England could show ?   Need we wonder, then, that the necessity for a new horse became only too apparent ! It was but a low-priced hack-hunter this time, yet a service-able animal, which carried its rider to probably as many meets as took place that winter within access of Nursted. And not that winter only, but the summer following, went past without apparently any further action in the way of carrying out the projected scientific programme.   We find the retired sportsman sojourning for a long time in the south of Scotland during that summer, visiting friends, shooting, and in short living as much after the old fashion as if he had never seen Davy at Rokeby, and no visions of chemistry lectures had ever floated before him.

But the momentous epoch of his life was now fast ap-proaching. This summer of 1824 saw the last of his rambles wherein the rocks around him made no direct and urgent appeal to him. Henceforth he was to have an occupation even more absorbing than any which had yet held him in thrall, and into this new employment he was to carry all the energy which had hitherto marked his doings in other pursuits.

CHAPTER VII.

At last Murchison had found a calling wherein his love of out-of-door life, and his inclination towards an intellectual employment of some sort, could find fitting scope. From this time forward it was to be his good fortune to have one engrossing occupation, which, while furnishing abundant exercise and amusement, should ere long enable him to make his name a kind of household word among geologists in every part of the world.

How it came about that a man with no previous scientific training should have been able to gain such a reputation, and gain it so rapidly, deserves our consideration. We might conjecture either that the science could have been no very recondite matter, or that the man must have been possessed of very extraordinary powers. Neither supposition would be quite just. Such was the state of geological science at the time, that a great work could be done by a man with a quick eye, a good judgment, a clear notion of what had already been accomplished, and a stout pair of legs.

It is of importance that the reader should see how this

came to be the case, in order that he may adequately realize
what Murchison's life-work actually was. I would ask him,
therefore, to accompany me in a necessarily brief survey of
the condition of geology in this country during the first
quarter of this century, with a glance at some of the
more salient characteristics of the leading geologists among
whom the retired captain and fox-hunter was now to take
his place. We shall in this way be enabled to follow more
definitely the kind of work which lay open to his hand, and
to note what incentives and obstacles surrounded him on his
entry upon this new career.

Looking back to the beginning of this century, we see
the geologists of Britain divided into two hostile camps, who
waged against each other a keen and even an embittered
warfare. On the one hand were the followers of Hutton
of Edinburgh, called from him Huttonians, sometimes also
Vulcanists or Plutonists; on the other, the disciples of
Werner of Freiberg, in Saxony, who went by the name of
Wernerians, or Neptunists. The strife lasted almost up to
Murchison's time, though it had in its last years waxed
faint and fitful. But many of the combatants who had been
in the thick of the fight were still alive when he assumed
the title of·geologist, and the current of geological thought
at that time had been largely influenced by the contest.

The Huttonians, who adhered to the principles laid down
by their great founder, maintained, as their fundamental
doctrine, that the past history of our planet is to be explained by what we can learn of the economy of nature at
the present time. Unlike the cosmogonists, they did not
trouble themselves with what was the first condition of the
earth, nor try to trace every subsequent phase of its history.

They held that the geological record does not go back to the beginning, and that therefore any attempt to trace that beginning from geological evidence was vain. Most strongly, too, did they protest against the introduction of causes which could not be shown to be a part of the present economy. They never wearied of insisting, that to the every-day workings of Air, Earth, and Sea must be our appeal for an explanation of the older revolutions of the globe. The fall of rain, the flow of rivers, the dash of waves, the slowly-crumbling decay of mountain, valley, and shore, were one by one summoned as witnesses to bear testimony to the manner in which the most stupendous geological changes are slowly and silently brought about. The waste of the land, which they traced everywhere, was found to give birth to soil— renovation of the surface thus springing Phœnix-like out of its decay. In the descent of water from the clouds to the mountains, and from the mountains to the sea, they recognised the power by which valleys are carved out of the land, and by which also the materials worn from the land are carried out to the sea, there to be gathered into solid stone—the framework of new continents. In the rocks of the hills and valleys they recognised abundantly the traces of old sea-bottoms. They stoutly maintained that these old sea-bottoms had been raised up into dry land from time to time by the powerful action of the same internal heat to which volcanoes owe their birth, and they pointed to the way in which granite and other crystalline rocks occur as convincing evidence of the extent to which the solid earth had been altered and upheaved by the action of these subterranean fires.

That a theory in many respects so bold and original, and

To face page 98.

JAMES HUTTON, M.D.
From an Original Portrait by Sir Henry Raeburn, in the possession of
Sir George Warrender, Bart.

embracing so wide a view of the whole field of inorganic nature, should be imperfect; that the full meaning of parts of it should not even have been suspected by its founder; that some of its details should have been built upon erroneous observations or deductions, may be readily believed. The most obvious imperfection about the theory was, that it took no account of the fossil remains of plants and animals. Hence it ignored the long succession of life upon the earth, which those remains have since made known, as well as the evidence thereby obtainable as to the nature and order of physical changes, such as alternations of land and sea, revolutions of climate, and such-like. But though the discovery of these profoundly significant truths opened up a world of research of which neither Hutton nor his friends had ever dreamed, it did not overturn what he had done. He had laid down principles which, in so far as they went, were true, and which the experience of successive generations has amply illustrated and confirmed. He had traced a bold outline which has been gradually filled in, but his master lines are traceable still. The whole of modern geology bears witness to the influence of the Huttonian school.

It was while views of this broad and suggestive nature were making way in this country, that others of a very different stamp came over from Germany. Werner at that time was teaching mineralogy at Freiberg, but he aspired to connect his science with a wide subject, and from the study of minerals to rise to the origin of the globe itself. He had not travelled. He had seen only a small corner of Europe, and having satisfied himself of the order and history of the rocks in that limited district, he proceeded to account for the formation of the various rocks of the rest of the globe

on the model of his own little kingdom.   Instead of starting
from what can be seen and known as to nature's operations
at the present time, Werner, like other cosmogonists, conceived himself bound to begin at the beginning.   He supposed that the earth had been originally covered with the
ocean, in which the materials of the minerals were dissolved.
Out of this ocean he conceived that the various rocks were
precipitated in the same order in which he found those of
Saxony to lie ; hence, on the retirement of the ocean, certain
universal formations spread over all the globe, and assumed
at the surface various irregular shapes as they consolidated.

Werner was a good mineralogist, and, as he classed rocks
by their mineral characters, there was a certain neatness
and precision about his system, and a facility of applying it
in other countries, such as no previous cosmological theory
could boast.   Moreover, as men were mineralogists before
geology came into existence, and as the general mineralogical
bias still prevailed, the doctrines of Werner, so largely based
on mineralogical considerations, had a great advantage in
the readiness with which they might be expected to be
adopted.   But, besides this, although his views about universal formations and the aqueous origin of all rocks—even
of basalt—were quite erroneous, he had grasped part of a
great truth in his chronological grouping of strata.   He
had likewise noticed, as indeed had been already to some
extent recognised by observers both in France and Germany,
that the remains of plants and animals imbedded in the
strata became fewer in number, and more unlike living
forms, the older the rocks in which they occur.   Even,
therefore, had he not been so full of zeal and eloquence as
to inspire his pupils with enthusiasm, his views would pro-

bably have made a way for themselves in Europe. But his ardour kindled a like spirit in those who came to listen to him. They returned to their own homes eager to apply, even in the most distant corners of the globe, the system which had been made so clear to them at Freiberg. They had at heart not only the cause of truth, but the fame of an eloquent teacher and friend, so that their course, at least in this country, became a kind of propagandism.

It is hardly possible now to realize how fierce and personal was the Huttonian and Wernerian war. Hutton himself had lived and died in Edinburgh. The crags and ravines of that romantic town had inspired him with some of his views, and, after he had gone, these features remained as memorials of his teaching, to friends who loved and followers who revered him. Edinburgh was naturally therefore the home of the Huttonian theory. It so happened, however, that in the year 1804 the Professorship of Natural History was given to Robert Jameson,—a student from Freiberg, full of the true Wernerian ardour. He was not long in office before he began to gather round him a band of disciples; and thus Edinburgh became a chief focus of the geological war.[1]

Amid the turmoil of the contest one figure still stands out prominently, calm and gentle, full of the courtesy of the days of chivalry, fighting not for self nor for fame, but generously setting lance in rest for the cause of truth, and on behalf of a revered teacher and friend—formidable in the lists withal, well skilled in defence, and with keen eye and ready hand to mark the weak points in his adversary's

[1] Among recently published reminiscences of this time, reference may be made to Sir Henry Holland's interesting allusions to the fierceness of the contest in Edinburgh.—See his *Recollections of Past Life*, p. 81.

armour. Such was the illustrious Playfair—a man to whom geologists owe a debt of gratitude which has perhaps never yet been adequately paid. Hutton had passed away, his work unfinished, and the style of his writings so obscure as to set a barrier to the general diffusion which their genius merited. Playfair, who was his warm personal friend, determined to prevent the risk of such doctrines as those of Hutton sinking into neglect, and to that end composed, and, in the spring of 1802, published his *Illustrations of the Huttonian Theory*.

This great work may be taken as the text-book of the Huttonian school. It contains not only the views taught by Hutton himself, but the expansion and application of them by Playfair. Gifted with an eloquence which, for dignity, precision, and elegance, reminds us of some of the best old French models, and which has certainly never since been equalled in the geological literature of this country, Playfair not only gained for the doctrines of his master a publicity and measure of acceptance which they might not otherwise have attained, but he raised geology out of the region of mere wild speculation, and placed it in an honourable position among the inductive sciences. The real rise of geology in this country into the dignity of a science, is traceable mainly to the influence of the *Illustrations of the Huttonian Theory*.[1]

But in the earlier years of the century this was not re-

[1] This was acknowledged six-and-twenty years after the *Illustrations* had appeared, and when their author had gone over to the majority. A warm and graceful tribute to his influence, with a frank recognition of the obligations of geologists to his labours in their service, was then given by Dr. Fitton in his Presidential Address to the Geological Society of London.—*Proc. Geol. Soc.* i. 56 (15th February 1828).

cognised. The very principles of geology were still matter of discussion. These would doubtless have been sooner settled but for the baneful influence of Wernerianism, and the check given by that system to the development of the views of Hutton and Playfair. Still it must in fairness be acknowledged, that Wernerianism introduced a more precise mineralogy and petrography than had ever been known before, and that though this was at the best but a poor substitute for the earlier growth of sound geology, it was an advantage, the loss of which, when it died out with that system, has in one not unimportant branch crippled British geology ever since.

In the midst of this ferment of conflicting theories, a few men interested in inquiries as to the nature and origin of minerals and rocks, drew together in London in the year 1807, and formed themselves into the Geological Society. A further reference to this important event will be made in the next Chapter, when we come to the time when Murchison joined the Society. In the meantime we may note that the aim of the founders was to gather facts as to the composition and structure of the earth without reference to questions of theory. With this view they met at short intervals to read papers on the rocks or minerals of particular species, or of special districts, and every few years gathered the more important of these papers into a large quarto volume of Transactions. During the early days of its existence the Society devoted itself with praiseworthy diligence to questions of mineralogy, or of the geological structure of different localities. The members hardly ever meddled with the remains of the plants and animals imbedded in the rocks. That these remains had a deep meaning, that they were to furnish the

key which would make it easy to group the rocks of England in their order of formation, and that they contained the
records of a marvellous march of life upon the earth, had not
yet dawned upon the minds of any of these early pioneers.

While the acknowledged leaders of the infant science of
geology were on the one hand wrangling as to the principles
to be adopted, and, on the other hand, busying themselves
with the collection and discussion of details of no great
moment, a man had been quietly and unobserved at work
for long years among the rocks of England, and had learned
their secret as none else had done. Born in Oxfordshire,
William Smith had been used in childhood to collect and
wonder over the fossils so abundant round his birthplace.
In later years, trained to the profession of civil engineer
and land-surveyor, he had recognised his early playthings in
far distant parts of the country. Step by step he was led to
perceive, in a far more precise and accurate way than had
been thought of by Werner or any previous observer, that
each group of strata had its own characteristic fossils. By
this test he could recognise a series of rocks all the way from
the coasts of Dorset to those of Yorkshire. He surprised
some of his friends who had made collections of fossils
by telling them from what special set of rocks each series
of shells had been obtained. He constructed, and as far
back as 1799 began to publish, geological maps of various
parts of England, on which the different groups of rocks
which he had made up were delineated with singular accuracy. At agricultural meetings, and to any inquirer who
wished to see them, he exhibited these maps, showing more
particularly their value in questions of farming and water-
supply. He had tried to find patrons, with whose help he

might publish a general work, and even issued prospectuses of his proposal, but failed to succeed, until at last, in the year 1815, he gave to the world his " Map of the Strata of England and Wales." But long before the appearance of this map, and of the other works which he issued in succession, his ideas had spread widely through the country. Hence when these marvellous productions were published, they met with immediate acceptance. They completely revolutionized the geology of the day, and called forth from his contemporaries the most unqualified praise, and the well-merited title of the Father of English Geology. It was now possible to arrange the rocks of the country in definite chronological order, to compare those of one district with those of another, to trace the connexion of the varying character of the strata underneath with the change of soils and the rise of springs. But, above all, William Smith's discoveries led the way to all that has since been done in tracing back the history of Life into the dim past. He was not himself a naturalist, but he laid that sure foundation on which our knowledge is built of the grand succession of living beings upon the surface of our planet.

From the prodigious impetus given by these revelations Geology made a new start in England, and branched out especially in two directions, which have continued up to the present time to be the paths chiefly followed by geologists in this country. In the first place, what is called Stratigraphical geology, that is, the accurate grouping of the rocks according to their order of formation, took its rise from the work of William Smith. Before his day no means existed of making any such subdivision beyond the vague general distinctions implied in such terms as Primary, Transition, and Secondary.

In the second place, from the attention now given to fossils as a key to the discrimination of rocks, the science of Palæontology, or the study of ancient forms of life, first took root in England. It is true that the researches of Cuvier among the extinct mammals of the Paris basin, and his clear and eloquent writings, as well as the labours of Brongniart, had drawn the eyes of the world to the interest attaching to fossil remains. These discoveries undoubtedly laid the foundations of Palæontology. They were not made, however, until after Smith's views, unpublished indeed, but freely communicated, had begun to spread in this country, and until consequently the minds of geologists were in some degree prepared for them by learning that a new meaning and value had begun to be discernible in the remains of the plants and animals imbedded in the rocks.

At the same time that this new development of geological inquiry took place, certain other changes came about in England. Foremost among these was the decay of Mineralogy and Petrography, or Mineralogical geology. Men found such a great untrodden field opening out before them, that they forsook the old and well-beaten paths of mineralogy. Neglecting the study of minerals, they left off also that of the mineralogical composition of rocks. For somewhere about half a century these branches of geology remained scarcely cultivated at all in this country, and only within the last few years have some of our geologists wakened up to the fact, that in this department of their science they have been far outstripped by their brethren of the hammer in Germany and in France.

So strongly did the tide now set in towards stratigraphical and palæontological pursuits, that another not less important

branch of geological research, which had been begun with
much promise, fell into neglect—the application of physical
experiment to the elucidation of geological problems.   The
merit of having started this line of investigation belongs to
the Huttonian school.   Among Hutton's friends and admirers was Sir James Hall of Dunglass—a man of singular
shrewdness, with a strong bent towards putting things to
the test of experiment, and an inventive faculty of no common order.   He had urged Hutton to apply this test to
some of his views which had been most keenly controverted.
That philosopher, however, had a deep conviction that as
we could never hope to imitate the scale of nature's operations, so we might run a great risk of having false impressions given to our minds by such experiments.   He seems
to have had a kind of contempt for those who "judge of the
great operations of the mineral kingdom from having kindled
a fire and looked into the bottom of a little crucible." [1]

Hall, though, from deference to his master, he generously
refrained from putting his ideas into practice during the
lifetime of the latter, felt sure that some parts of the Huttonian Theory could be proved or disproved by simple experiments.[2]   After Hutton's death a series of trials, memorable as the birth of Experimental Geology, proved the truth
of his surmise, adding, at the same time, to the stability
of Hutton's views and the fame of the Scottish School of
Geology.   During the first quarter of this century he published at intervals a series of admirable papers in the Transactions of the Royal Society of Edinburgh on such questions
as the igneous origin of basalt-rocks, the formation of marble

[1] *Theory of the Earth,* vol. i. p. 251.
[2] *Transactions of the Royal Society, Edinburgh,* vol. vi. p. 76.

and crystalline limestone, the contortion of the earth's crust, etc. But the vitality of the geological school in that capital was gone. No one followed up the path opened by Hall, and men were too busy elsewhere in making out the order of the rocks and the succession of the fossils to have time or inclination for theoretical questions.

Another change of import in the history of the time succeeding the publications of William Smith, was the gradual decline and extinction of Wernerianism. Even at its stronghold in Edinburgh· it had been waning. Two months after the founding of the Geological Society of London, Jameson had started the Wernerian Society in Edinburgh—a Society which continued for many years, in spite of its name, to do much excellent work in various departments of natural history. Its founder had come to be regarded as the avowed leader of the Wernerians of this country. He had one great advantage over his opponents. Accurate mineralogical knowledge enabled him to discriminate rocks with a precision to which they could make no pretension, and although this was an accomplishment of little real moment in the theoretical questions chiefly in dispute, he did not fail to make the most of it, nor they to betray their consciousness of their inferiority in that respect. In the end, however, Jameson and his band of co-believers in Werner came to be gradually isolated on the rocks of Edinburgh with an ever-rising flood of the dominant geology around them. There they stood, battling as well as they might with the inevitable, until at last Jameson frankly acknowledged, at one of the evening discussions of the Royal Society, that Wernerianism was doomed and deserved to die.[1]

[1] This incident, of Jameson's confession, was told to the writer by Sir

PROFESSOR ROBERT JAMESON.
From a Miniature in the possession of Dr. Claud Muirhead, Edinburgh.

It had been one of the characteristics of Werner's system to ignore, or at least to neglect, volcanoes and volcanic action. There were no volcanoes in that little kingdom which he had taken as the model of the globe. The neglect was pardonable perhaps in his case, but when his votaries in travelling over the world met face to face with only too manifest proofs of the vitality of the internal heat of the earth, they had recourse to every possible explanation—the combustion of subterranean beds of coal, or indeed any supposition that would depreciate the importance of volcanoes as parts of the general economy of the world. They almost seemed to regard volcanoes with dislike as anomalous interferences with the normal constitution of things. They denied the igneous origin of such rocks as basalt, even though their opponents proved that rocks of precisely similar character had often been seen flowing in a melted state down the sides of volcanoes. Excellent service had been done in exposing the absurdity of these notions by Desmarest, Montlosier, Faujas St. Fond, and other geologists on the Continent, and in this country by Macculloch, Boué, and others, but by none more signally than by Mr. Poulett Scrope in his admirable memoirs on the volcanic districts of Naples and Central France, and his work on Volcanoes.[1] Though the British Islands abound in

Robert Christison, and by Professor Balfour, who were present at the Royal Society of Edinburgh when it took place. It has not been possible to recover the date of the meeting.

[1] Mr. Scrope, to whose cordial friendship it is a pleasure to record my obligations, has furnished me with the following reminiscence of this early geological controversy :—" I well recollect, on a discussion arising at the Geological Society meeting, after the reading of a paper of mine on the Auvergne volcanoes, Greenough's arguing that the cinder-cones *might* be volcanic, but that the plateaux of basalt that adjoined them were certainly precipitations from the archaic ocean of Werner. In my reply I got the laugh in my favour by putting to him whether, if he found

old volcanic rocks of many ages, even the stimulus of the success of these observers was not enough to divert a few able observers from the general drift of English geology into the channels already indicated. Another generation had to arise before the volcanic history of Britain began to attract serious notice.

But of the changes which followed the rise and rapid development of stratigraphical geology and palæontology in England, perhaps the most regrettable was the neglect of what is now termed Physiographical Geology—that is, the study of the origin of the present external features of the land. Hutton and Playfair were full of this subject. They refused to admit of hypothetical revolutions, but steadfastly insisted on explaining the changes of the past by the same kinds of action which may still be seen at work. Nevertheless, though they directed attention so forcibly to the every-day changes of the earth's surface, their teaching did not displace the more sensational hypotheses of catastrophe and convulsion. It was reserved for a foreign scientific Society to recall the thoughts of men to the revolutions which the land had undergone within the time of human chronicles, and for the illustrious Von Hoff to gather the historical evidence of these revolutions[1]—a task which has since been so worthily followed up and extended by Lyell.

some morning on entering his library (so well known to geologists through his hospitalities) a stream of ink flooding the carpet, with a broken bottle at one end of it, he would be satisfied with the explanation of his housemaid that she had broken the bottle, but was wholly innocent of spilling the ink, which must have been done in some other way and at some other time. Greenough lived and, I believe, died a consistent Wernerian, and many a contest I had with him in 1823-5 on the identical character of basalt and lava."

[1] Geschichte der durch Ueberlieferung nachgewiesenen natürlichen Veränderungen der Erdoberfläche—ein Versuch von K. E. A. von Hoff.

While dwelling on ordinary and familiar agents of change, the Scottish philosophers found in these the explanation of the origin of much of the scenery of the land. They delighted to trace the origin of valleys, the sculpturing of mountains, and the gradual evolution of the various features of a landscape. They attributed these irregularities of surface to the action of rain and streams upon masses of land upheaved above the sea ;—an idea which was deemed too bold even by many of their boldest followers, such as Hall, and which fell into comparative oblivion. It was noticed in text-books and treatises only to be dismissed as extravagant. In its place the notion prevailed that to subterranean action we must mainly attribute the present inequalities of the land—a notion which has been prevalent until within the last few years, when the rising generation of geologists has begun to recognise the true meaning and place of the Huttonian doctrine. We shall find that Sir Roderick Murchison up to the close of his life battled for the supremacy of the underground forces in the modelling of the surface of the land. And yet he had read the lucid observations and arguments of Mr. Poulett Scrope, written so far back as 1822, to prove how valleys in central France had been cut out by running water ;—nay, as we shall see, one of his earliest geological tours abroad was to this very region, where he became convinced of the truth of Mr. Scrope's views, though the conversion proved to be only a transient one.

In fine, the first quarter of the present century was a time of marvellous vigour in the history of geology. It was during

This work was begun as a prize-essay written in response to an invitation from the Royal Society of Sciences at Göttingen. The first volume was published in 1822.

that time that the science took shape and dignity. Amid the conflict of opposing schools progress had been steady and rapid. Every year broadened the base on which the infant science was being built up. The rocks of England and Wales were arranged in their order of age, the outlines traced by Smith having been more and more filled in. Excellent service had been done by the admirable handbook of Conybeare and Phillips, while Boué, Jameson, Macculloch, and others, had made known the rocks of large tracts of Scotland. But a vast deal remained to be accomplished. The field was still in a sense newly discovered, it stretched over a wide area, and lay open to any one who with active feet, good eyes, and shrewd head chose to enter it. And the enthusiasm of those who were already at work within its borders sufficed not only to inspirit their own labours, but to attract and stimulate other fellow-workers from the outer world.

From the foregoing rapid survey of the progress of geology during the first quarter of this century we can see the probable line of inquiry which any young Englishman would then be likely to take who entered upon the pursuit of the science without being gradually led up to it by previous and special studies. In the first place, he would almost certainly be a Huttonian, though doubtless holding some of Hutton's views with a difference. He would hardly be likely to show much sympathy with the fading dogmas of the Wernerians.

In the second place, he would probably depart widely from one aspect of the original Huttonian school in avoiding theoretical questions, and sticking, possibly with even too great pertinacity, to the observation and accumulation of facts.

In the third place, he would most likely have no taste for experimental research as elucidating geological questions, and might set little store by the contributions made by physicists to the solution of problems in his science.

In the fourth place, he would almost certainly be ignorant of mineralogy, and whenever his work lay among crystalline rocks it would be sure to bear witness to this ignorance.

In the fifth place, devoting himself to what lies beneath the surface as the true end and aim of geology, he would be apt to neglect the study of the external features of the land. And this neglect might lead him in the end to form most erroneous views as to the origin of these features.

Lastly, his main geological idea would probably be to make out the order of succession among the rocks of his own country, to collect their fossils, unravel their complicated structure, and gather materials for comparing them with the rocks of other countries. In a word, he would in all likelihood drift with the prevailing current of geological inquiry at the time, and become a stratigraphical geologist.

There was no reason in Murchison's case why these influences of the day should not mould the whole character of his scientific life. We shall trace in the records of later years how thoroughly they did so. As he started, so he continued up to the end, manifesting throughout his career the permanent sway of the circumstances under which he broke ground as a geologist.

At first the novelty and fascination of the pursuit engaged his attention. Many a time on his walking and hunting expeditions he had noticed marine shells far inland. He now found out that such shells formed, as it were, the

alphabet of a new language, and that by their means he
might decipher for himself the history of the rocks with
whose external forms he was so familiar.   He threw himself
into the study with all his usual ardour, and ere long
became as enthusiastic with his hammer over down and
shore as he had been with his pencil and note-book among
the galleries of Italy, or with his hunting-whip or his gun
across the moors of Durham.

Of the men on whom the progress of geology mainly
depended at the time when Murchison joined them to
become their life-long associate and friend, something should
be said here.   Some of the band of enthusiasts by whom the
Geological Society of London was originated still lived, and
took an active share in the Society's work.   Among them
were Greenough, the true founder and first president of the
Society—amiable, yet shy, and somewhat hesitating in
manner, full of all kinds of miscellaneous knowledge, obsti-
nately sceptical of new opinions, a kind of staunch geological Tory, and playing the part of objector-general at the
evening discussions ; and Babington, a kindly, bland, and
courteous veteran, who, well versed in the mineralogy of his
time, had gathered at his house the few like-minded friends
from whom the Geological Society sprang, who introduced the
practice of discussing the papers read at the meetings, and
who even when nearly fourscore years of age found a congenial occupation in the Society's museum.   Other names
which had long been associated with the progress of the
Society still had an honoured place on its list.   Such were
those of Wollaston—admirable mineralogist, sternly upright
in his search for truth, quiet, reserved, serious, looking like a
Greek sage, and deservedly regarded as a general arbiter in
the scientific world of London, yet, to those who were privi-

To face page 115.

REV. WILLIAM D. CONYBEARE, F.R.S.
From a Photograph in the possession of his Family.

leged with his more intimate friendship, fond of a joke and
of a quiet corner in a pheasant cover, where his gun seldom
failed to tell; Warburton—cautious and uncommunicative;
Fitton—friendly and painstaking, an active leader in the
affairs of the Society, but somewhat hasty in temper, and
prone to what some of his colleagues thought "red-tape"
formality, yet an admirable observer in the field, a most
gifted debater, and one whose clear and elegant pen did good
service to the infant science in popular journals, and whose
house formed a pleasant centre for the geologists of town;
Conybeare—clear-headed, critical, full of quaint humour and
wit; Buckland—cheery, humorous, bustling, full of eloquence,
with which he too blended much true wit, seldom without his
famous blue bag, whence, even at fashionable evening parties,
he would bring out and describe with infinite drollery, amid
the surprise and laughter of his audience, the last "find" from
a bone-cave; Leonard Horner—mild, unpretending, and defer-
ential, yet shrewd and systematic, a valuable member of the
council of management of the Society; Sedgwick—with his
well-remembered hard-featured yet noble face, and eyes like
an eagle's, manly alike in body and mind, full of enthusiasm,
ready and graphic in talk, generous and sympathetic, often
depressed by a constitutional tendency to hypochondria, yet,
when in full vigour of health, shrinking from no toil, either
at home or abroad, in furtherance of his chosen branch of
science, and laying up year by year a store of facts and of
brilliant deductions from them, which have given him one
of the most honoured places in the literature of geology.

Later in advent than these magnates, or less prominent at
the time with which we are now dealing, and therefore more
of the standing of Murchison himself, came Lyell (now a household name all over the world), even then noted among his

fellows for those qualities the further development of which has been of such value to the spread of sound geology, and specially for his earnest pursuit of information on every subject which could throw any light upon the problems of the science ; Henry De la Beche, then a handsome and fashionable young man, just beginning to show that quick and shrewd observation of nature, and rare power of philosophical induction which eventually gave him so honourable a rank in British geology ; Dr. Edward Turner—young, open, unassuming, but eager in quest of knowledge, and one of the first chemists to recognise the necessity of linking chemistry closely with mathematics ; G. Poulett Scrope—full of geological zeal, which led him through the chief volcanic districts of Europe, and stimulated him to produce an early series of writings which the avocations of a subsequent political life have left all too few ; W. J. Broderip— active and methodical, full of varied natural-history knowledge, brimming with joke, yet taking a keen interest in the affairs of the Society, and keeping them in order, not with the severe rigour of Dr. Fitton, but with an easy good-humoured precision which pleased everybody and did the Society and its members most excellent service.

Many other names of not less note should receive more than passing mention here among Murchison's early scientific contemporaries. Such were Whewell, Herschel, C. Stokes, Babbage, Webster, Lonsdale, Sir Philip Egerton, the Earl of Enniskillen (then Viscount Cole), and others, most of whom have passed away. Some of these men became intimate personal friends of the subject of this biography, and their names will therefore appear frequently in the subsequent chapters.

CHAPTER VIII.

WE return to Murchison's career. He had now fairly resolved to cast in his lot with the men of science. Bringing his wife with him from Nursted, he came up to London, and rented the house No. 1 Montague Place, Montague Square. There, settling down to a much more serious employment than any he had yet undertaken, he entered upon his new life full of ardour and hope.

" If in the last years of my fox-hunting," he says, " I began to sniff up a little scientific knowledge, and showed a willingness to turn my former rambles among the Alps and Apennines to some profit, it was only in the winter of 1824 that I buckled resolutely to the study of chemistry and the cognate subjects by attending Brande's early morning lectures at the Royal Institution. This I did by the advice of Sir H. Davy as a necessary preliminary. From this moment, all horses except a pair for my wife's carriage being dismissed, I got quite into another and to me an entirely new phase of society. My note-books chiefly refer, however, to the geological lectures, and before the spring came I became

acquainted, through books and lectures, with the chief phenomena of British geology. Though chemistry never had
strong attractions for me, I kept regular notes of the lectures
on its various branches, and, at the end of my course, knew
as much about that science as was necessary for a field-
geologist." [1]

In later years he used to recall with no little pleasure an
incident in that course of lectures.   One day Dr. Brande did
not lecture, and his place was taken by his assistant—a
pale thin lad, who began with some timidity, but gathering
courage as he went on, soon proved himself to be an admirable lecturer, and received from his delighted audience a
hearty round of applause.   It was Michael Faraday.[2]

From the Royal Institution lectures the transition was
easy to the papers and debates to be heard in those little
rooms in Bedford Street, Covent Garden, where the Geological
Society then held its meetings.   We have in the preceding
chapter noticed the place which the creation of this Society
fills in the history of geological science in this country.
Some further details of a more personal kind may here be
given, partly because the men who started the Society were
in great measure still living and active members of it when
Murchison joined them, partly because Murchison's own
scientific career was closely bound up with the subsequent
history of the Society, and partly because the work done by

[1] These notes, which still exist, show a vast deal of diligence, and a
very fair amount of knowledge. They seem to have been carefully
written out from day to day, and with equal fulness, whether the subject
of the lectures was the composition of beef or the properties of oxygen.

[2] In telling this story to the writer only a few months before his death,
Murchison said it was Faraday's first lecture. A comparison of dates,
however, shows that his memory had been at fault, for Faraday had
already gained a reputation as experimenter and original investigator
before this time.

the Society, and its influence upon the progress of the science, have been so great that they claim grateful recognition, and deserve adequate record in any work which professes to sketch, even in outline, the growth of a portion of British geology.

At the beginning of this century original research in natural science was promoted in London by two Societies, the Royal and the Linnean. Next in order of time came the Geological Society, which took its origin, as already mentioned, in 1807, and under the following circumstances :[1]—

" Count de Bournon had written an elaborate monograph on carbonate of lime, and, in order to raise funds for its publication, Dr. Babington invited to his house a number of gentlemen distinguished for their zeal in mineralogical knowledge, when a subscription-list was opened, and the necessary sum was collected. Other meetings of the same gentlemen took place for friendly intercourse, and it was then proposed to form a Geological Society. On November 13, 1807, a meeting was held at the Freemasons' Tavern, Great Queen Street, at which resolutions were passed formally constituting the Society. Only eleven gentlemen were present, and their names deserve to be recorded. They were Arthur Aikin, William Allen, F.R.S., William Babington, M.D., F.R.S., Count Bournon, F.R.S., H. Davy, Sec.R.S.; J. Franck, M.D., G. Bellas Greenough, M.P., F.R.S., R.

[1] This narrative is taken from an account of the Society written by one of its Fellows, Mr. W. S. Mitchell, just previous to its recent change of quarters to Burlington House, and published in *The Hour* of November 5th, 1873. It is the only narrative which has been published of the early struggles of the Society. Compiled from the minute-books of the Society, it presents a reliable account of events which must always have an interest for English geologists.

Knight, J. Laird, M.D., J. Parkinson, Richard Phillips. Two
other supporters of the scheme, W. H. Pepys and William
Phillips, were unavoidably prevented from attending the
meeting, but their names were added to the list. The thirteen names were read out, and these gentlemen constituted
themselves the first members of the Geological Society, with
the resolution,—' That there be forthwith instituted a Geological Society for the purpose of making geologists acquainted
with each other, of stimulating their zeal, of inducing them
to adopt one nomenclature, of facilitating the communication
of new facts, and of ascertaining what is known in their
science, and what remains to be discovered.'

" The customs of the new association were such that it
would now be called a Club rather than a Society. The
members were to meet on the first Friday of every month at
five o'clock, at the Freemasons' Tavern, for a fifteen shilling
dinner. Business was to commence at seven o'clock, and
the chairman was to leave the chair at nine."

After drawing up rules and other initial formalities, including the election of a Patron (Right Honourable Charles
F. Greville, F.R.S.) and a President (G. B. Greenough, M.P.,
F.R.S.), the members, in accordance with one of their laws,
set themselves to work in " contributing to the advancement
of geological science, more particularly as connected with
the mineral history of the earth." Their numbers increased,
and among their early adherents they could count even the
President of the Royal Society, who requested admission into
their ranks. Specimens of minerals were presented to them
with such liberality that within a year the idea took definite
shape of securing some permanent place for the collections
and meetings of the Society. Accordingly, in 1809, rooms

were obtained at No. 4 Garden Court, Temple, and there the infant Society was enabled first to erect its household gods. But this step, so indicative of independence and activity, soon led to serious troubles.

" The Society reckoned among its members many who were Fellows of the Royal Society, and so long as it aimed at nothing more than dining once a month and discussing geological subjects, there was nothing to which the Fellows of the Royal Society could raise any objection. But as soon as a separate habitation was proposed, with a separate collection of specimens, it was at once objected that the dignity of the Royal Society would be impaired. At the meeting on March 3 (1809), Sir Joseph Banks sent in his resignation, and soon after a proposal was made by the Patron, the Right Hon. Charles Greville, to make the Geological Society an assistant association to the Royal Society. The drift of the plan was, that the Geological Society should consist of two classes of members—(1.) those who were Fellows of the Royal Society, and (2.) those who were not. That all papers should be sent to the Royal Society for them to select what they liked for publication, and that the Geological Society should be at liberty to publish the rejected papers if they wished. A special meeting to consider this proposal was held at the Freemasons' Tavern on March 10, when this resolution was passed :—' That any proposition tending to render this Society dependent upon or subservient to any other Society does not correspond with the conception this meeting entertains of the original principles upon which the Geological Society was founded.' The proposal was decided to be inadmissible, and it was pointed out that it was never intended to impose any obligations on members of the Geo-

logical Society inconsistent with their allegiance to the Royal
Society.   Mr. Greville sent in his resignation as Patron, but
the firmness shown by a few of the promoters of the Society
secured for it freedom and independence of action."

This vigorous action no doubt helped to strengthen the
Society both in numbers and in influence.   Even so early as
1810 the first habitation at the Temple was found too small,
and the chattels of the Society were in that year transferred
to No. 3 Lincoln's Inn Fields.   Further evidence of vigour
was shown by the fact that the papers read at the meetings
began in 1811 to be published in quarto volumes of the
massive orthodox size, and with wealth of margin and illustrations.   After six years of great activity, the need for
further space again became urgent.   Another migration
took place, the rooms selected being at No. 20 Bedford Street,
Covent Garden.   For twelve years, that is from 1816 to
1828, the Society continued to hold its meetings in that
building.   It was while there that " in 1825 a Charter of
Incorporation was applied for and obtained from George IV.,
the date of affixing the royal seal being April 23, ' in the
sixth year of our reign.'   The five members named in the
charter were,—W. Buckland, Arthur Aikin, John Bostock,
G. Bellas Greenough, and Henry Warburton.   Dr. Buckland
was by the charter appointed first President."

The Geological Society of London "was, in its early days,"
to quote the words of one of its former most distinguished
members, " composed of robust, joyous, and independent
spirits, who toiled well in the field, and did battle and cuffed
opinions with much spirit and great good will; for they had
one great object before them—the promotion of true knowledge—and not one of them was deeply committed to any

system of opinions." The same writer boasts of " the joyous meetings, and of the generous, unselfish, and truth-loving spirit that glowed throughout the whole body." [1]

It was into this pleasant gathering of enthusiasts that Murchison found his way in the winter of 1824-25. " I entered the Society," he says, " Professor Buckland of Oxford being President, and on the 7th of January took my seat, and had my hand shaken by that remarkable man, who was then giving such an impulse to our new science, and was of course my idol. One of the honorary secretaries, then a young lawyer, was Charles Lyell, who then read his first paper, on the marl-lake at Kinnordy, in Forfarshire, the property of his father.

" Among my scientific friends I was of course most proud to reckon Dr. Wollaston, who then and in subsequent years invariably took pains to make me understand the true method of searching after new facts, and often corrected my slips and mistakes.

" I also owed great obligation to Mr. Thomas Webster. His acquaintance with minerals and ores, as well as with fossil animal remains, and his well-composed descriptions, were strikingly illustrated by his great powers as an artist. Born in the Shetland Isles, and there receiving a good education, Webster had never seen in that region a tree higher than a bush, so that in coming southwards, as he told me, he never could forget the astonishment and admiration he felt, when on reaching the valley of Berriedale, on the borders of Sutherland, he for the first time saw true forest-trees. Before these he kneeled down, as true a worshipper as Linnæus when he first beheld in England the yellow blossom of our common furze.

[1] Sedgwick, *Brit. Pal. Fossils*, Introduction, pp. xc. xcii.

" Sedgwick, Whewell, Peacock, Babbage, Herschel, and all the eminent Cantabs of the time, came flocking in continually to our scientific assemblies. From his buoyant and cheerful nature, as well as from his flow of soul and eloquence, Sedgwick at once won my heart, and a year only was destined to elapse before we became coadjutors in a survey of the Highlands, and afterwards of various parts of the Continent."

To show further the contrast between his employments in London and his amusements during previous winters in the country, it may be well to note that he not merely made a good many acquaintances among scientific people, but became a personal friend of not a few men who then or afterwards stood in the foremost ranks of literature. He met Thomas Moore, Hallam, Copley (Lord Lyndhurst), Lord Dudley, and others, who used to frequent the soirées of Miss Lydia White, whose well-known ambition it was to gather round her the intellect and taste of London society.[1]

With lectures on science, scientific papers and discussions, evening soirées, and the opportunity of hearing and talking to men who had already made themselves famous, he found enough fully to fill up his time, and to make London life a very different thing to him from what it had been in the old days when he used to escape to town from the monotony of a country barrack. With his characteristic ardour, he had not completed his first winter's studies before he longed to be off into the field to observe for himself.

" My first real field work," he says, " began under Pro-

---

[1] Sir Walter Scott, who knew this lady well, describes her as " what Oxonians call a lioness of the first order, with stockings nineteen times nine dyed blue, very lively, very good-humoured, and extremely absurd." —*Life*, vol. ii. p. 137.

fessor Buckland, who having taken a fancy to me as one of
his apt scholars, invited me to visit him at Corpus Christi
College, Oxford, and attend one or two of his lectures. This
was my true launch. Travelling down with him in the Oxford coach, I learned a world of things before we reached the
Isis, and, amongst others, his lecture on Crustacea, given
whilst he pulled to pieces on his knees a cold crab bought at
a fishmonger's shop at Maidenhead, where he usually lunched
as the coach stopped.

" On repairing from the Star Inn to Buckland's domicile,
I never can forget the scene which awaited me. Having, by
direction of the janitor, climbed up a narrow staircase, I
entered a long corridor-like room (now all destroyed), which
was filled with rocks, shells, and bones in dire confusion,
and, in a sort of sanctum at the end, was my friend in his
black gown looking like a necromancer, sitting on the one
only rickety chair not covered with some fossils, and clean-
ing out a fossil bone from the matrix."

The few days at Oxford were memorably pleasant. Buckland's wit and enthusiasm glowed through all his scientific
sayings and doings, and he had a rare power of description
by which he could make even a dry enough subject fascinat-
ingly interesting. Murchison heard one or two brilliant
lectures from him, but what was of still more importance, he
accompanied the merry Professor and his students, mounted
on Oxford hacks, to Shotover Hill, and for the first time in
his life had a landscape geologically dissected before him.
From that eminence his eye was taught to recognise the
broader features of the succession of the oolitic rocks of England up to the far range of the Chalk Hills ; and this not in
a dull, text-book fashion, for Buckland, in luminous language,

brought the several elements of the landscape into connexion with each other and with a few fundamental principles which have determined the sculpturing of the earth's surface. His audience came to see merely a rich vale in the midst of fertile England, but before they quitted the ground the landscape had been made to yield up to them clear notions of the origin of springs and the principles of drainage.

This was the very kind of instruction needed to fan the growing flame of Murchison's zeal for science. He returned to town burning with desire to put his knowledge to some use by trying to imitate, no matter how feebly, the admirable way in which the Oxford Professor had applied the lessons of the lecture-room to the elucidation of the history of hills and valleys. While shooting and rambling, as he had so often done, at the house of his father-in-law, he had already noted many geological facts in the district around Petersfield without paying much heed to them, or seeking in any way for their explanation; but from what he had learnt from Mr. Webster and Dr. Fitton as to the Isle of Wight, he could see that in that island he should most likely find materials for understanding the geology of Petersfield. Accordingly he determined that this should be his first essay in independent field-work. Of this time he writes : " I was *totus in illis,* and making every preparation for a thorough survey of all the South coast—a project which was gladly backed up by my wife, who now saw that I was fairly bitten with my new hobby. Conybeare and Phillips' *Geology of England and Wales* had then become my scientific bible, and I saw that a fine field was opening for any zealous and active searcher after truth in completing many gaps which they had left to be filled up."

The summer of 1825 brought Murchison and his wife back once more to Nursted House, but the Hambledon fox-hounds had now lost their charms for him. With the same zeal he had thrown himself into another kind of hunting, in which, instead of old Parson Richards and his friends, he had for companion his own wife. Many a deep lane and rocky dingle did they explore together for fossils. Dr. Fitton came down to visit them and joined in the pursuit, tracing out by degrees the well-marked succession of cretaceous strata shown in that district.

Seeing in this way the problems which he had to work out in the Petersfield district, Murchison started with his wife in the middle of August on a tour of nine weeks along the South coast, from the Isle of Wight into Devon and Cornwall. Taking a light carriage and a pair of horses, he made the journey in short stages, lingering for days at some of the more interesting or important geological localities. Driving, boating, walking, or scrambling, the enthusiastic pair signalized their first geological tour by a formidable amount of bodily toil. Mrs. Murchison specially devoted herself to the collection of fossils, and to sketching the more striking geological features of the coast-line, while her hus-band would push on to make some long and laborious detour. In this way, while she remained quietly working at Lyme Regis, he struck westward for a fortnight into Devon and Cornwall, to make his first acquaintance with the rocks to which in after years Sedgwick and he were to give the name by which they are now recognised all over the world.

It was in the course of this tour that he met with a man whom he has the merit of having brought into notice, and who certainly amply requited him by the services rendered

in later years.   William Lonsdale had served in the Peninsular war, and retired on half-pay to Bath.   With the most
simple and abstemious habits his slender income sufficed not
only for his wants, but for the purchase of any book or fossil
he coveted, and so he spent his time in studying the organic
remains, and specially the fossil corals, to be found in his
neighbourhood.   Murchison met him accidentally in some
quarries,—" a tall, grave man, with a huge hammer on his
shoulder,"—and found him so full of information that he
stayed some days at Bath under Lonsdale's guidance.

With the enlargement of view which so instructive a
ramble had given him, Murchison prepared and read to the
Geological Society, on 16th December 1825, his first scientific
paper,—" A Geological Sketch of the North-western extremity of Sussex, and the adjoining parts of Hants and Surrey."[1]
This little essay bore manifest evidence of being the result
of careful observation of the order of succession of the rocks
in the field, followed by as ample examination of their
fossils as he could secure from those best qualified to give an
opinion upon them.   In these respects it was typical of all
his later work.

Having shown by this first publication his capacity as
an observer and describer, and being further recommended
by the leisure which his position of independence enabled
him to command, he was soon after elected one of the two
honorary secretaries of the Geological Society.   "Lyell being
then a law-student, with chambers in the Temple, could only
devote a portion of his time to our science, and was glad to
make way as secretary to one who, like myself, had nothing
else to do than think and dream of geology, and work hard
to get on in my new vocation."

[1] See *Geol. Trans.*, 2d ser., vol. ii. p. 97.

WILLIAM HYDE WOLLASTON, M.D.
From a Drawing by Sir Thomas Lawrence.

In the spring of 1826 he was elected into the Royal Society—an honour more easily won then than now, and for which, as the President, his old friend Sir Humphry Davy told him, he was indebted not to the amount or value of his scientific work, but to the fact that he was an independent gentleman having a taste for science, with plenty of time and enough of money to gratify it. His acquaintance with the scientific men of London daily increased, Davy and Wollaston being specially attentive in their encouragement. Of his intercourse with the latter he writes : " Wollaston's little dinners of four or five persons were most agreeable, and you were sure to come away with much fresh knowledge. A good dish of fish, a capital joint and some game, followed by his invariable large pudding, filled in with apples, apricots, or green-gages, all served on plain white porcelain by two tidy, handsome women, was the bill of fare.

" This was perhaps about the happiest period of my life. I had shaken off the vanities of the fashionable world to a good extent—was less anxious to know titled folks and leading sportsmen—was free of all the cares and expenses of a stable full of horses—and had taken to a career in which excitement in the field carried with it occupation, amusement, and possibly reputation."

But if distinction was to be won in this new kind of activity, it could only be by hard toil in the field. He had never had any of the special training which would have fitted him for working out geological problems indoors, such as the discrimination of fossils, or the characters and alterations of minerals and rocks ; hence, although stress of weather, not to speak of the pleasures of society, brought him to London and

kept him there during the winter and spring, he soon saw
that to insure progress in his adopted pursuit he must spend
as much as possible of every summer and autumn in original
field-exploration. He had begun well in this way by the tour
along the South coast. Now that another summer had come
round he prepared to resume his hammer in the field. As
before, a definite task was given to him. Buckland and
others advised him to go north and settle the geological age
of the Brora coal-field, in Sutherlandshire. Some geologists
maintained that the rocks of that district were merely a part
of the ordinary coal, or carboniferous system; others held
them to be greatly younger, to be indeed of the same general
age with the lower oolitic strata of Yorkshire. A good
observer might readily settle this question. Murchison
resolved to try.

Again he prepared himself by reading and study of fossils
to understand the evidence he was to collect and interpret;
and in order to do full justice to the Scottish tract, he went
first to the Yorkshire coast and made himself master of the
succession and leading characters of the rocks so admirably
displayed along that picturesque line of cliffs. The summer
had hardly begun before he and his wife broke up their
camp in London and were on the move northward.

At York he made the acquaintance of two men with
whom he was destined in after life to have much close intercourse and co-operation,—the Rev. William Vernon (afterwards Vernon Harcourt) and Mr. John Phillips. The latter
friend has kindly contributed the following reminiscences
of this interview:—" In a bright afternoon of early summer,
while engaged in museum arrangements, a man of cheerful
and distinguished aspect was presented to me by the Pre-

sident of the Yorkshire Philosophical Society, Mr. W. Vernon (Harcourt), as Mr. Murchison, a friend of Buckland, desirous of consulting our collections. The museum was tolerably well supplied with oolitic fossils, and especially those of the coralline oolite and calcareous grit of Yorkshire. Some of these were amusing enough. A diligent collector at Malton, who supplied the museum with specimens, sometimes brought what were called 'beetles,' made by painting and varnishing parts of shells and crustaceans. After examining the 'fossils' with care, Murchison *would* see these 'curiosities.' As it happened, they were laid contemptuously at the base of vertical cases, and were rather difficult to get out—'Never mind,' said the old soldier, 'we will lie down and reconnoitre on the floor.' I knew then that geology had gained a resolute disciple, possibly a master-workman."

Murchison's own record of the meeting is as follows :— " Phillips, then a youth, was engaged in arranging a small museum at York. He recommended me strongly to his uncle, William Smith, who was then living at Scarborough, and had little intercourse with the Geological Society, for he thought that Greenough and others, in taking from him the main materials of his original Geological Map of England, had done him an injustice. The unpretending country land-surveyor, who had really the highest merit of them all, had been somewhat snubbed by such men as Dr. Macculloch and others, who, having a superior acquaintance with the chemical composition of rocks and minerals, did not appreciate the broad views of Smith.

" From the moment I had my first walk with William Smith (then about sixty years old), I felt that he was just

the man after my own heart; and he, on his part, seeing
that I had, as he said, ' an eye for a country,' took to me
and gave me most valuable lessons. Thus he made me
thoroughly acquainted with all the strata north and south
of Scarborough. He afterwards accompanied me in a boat
all along the coast, stopping and sleeping at Robin Hood's
Bay. Not only did I then learn the exact position of the
beds of poor coal which crop out in that tract of the eastern
moorlands, but collecting with him the characteristic fossils
from the calcareous grit down to the lias, I saw how clearly
strata must alone be identified by their fossils, inasmuch as
here, instead of oolitic limestones like those of the south
we had sandstones, grits, and shales, which, though closely
resembling the beds of the old coal, were precise equivalents
of the oolitic series of the south. Smith walked stoutly
with me all under the cliffs, from Robin Hood's Bay to
Whitby, making me well note the characteristic fossils of
each formation."

Though the main object of this summer tour was to
work out the geological problem which had been assigned
to him in Sutherlandshire, he sketched a most circuitous
route, partly for the sake of showing Mrs. Murchison something more of the Highlands than she had yet seen, and
partly with the view of putting to use his new acquirements
in geology; so that after reaching Edinburgh, and having its
geology expounded to him by Jameson, instead of striking
north at once, he turned westwards to the island of Arran,
and spent many weeks among the Western islands, from the
Firth of Clyde to the north of Skye. The hills of his native
country had now acquired an interest for him which they
never possessed, even in the days when they drew him off

in eager pursuit of grouse and black-cock. At every halt his first anxiety was to know what the rocks of the place might be, and how far he could identify their geological position.   In Arran he filled his note-book with observations and queries about granite, red sandstone, limestone, and other puzzling matters, on which his previous experience in field-work in the south of England and in Yorkshire could throw no light, and for the elucidation of which he wisely resolved to secure at some future time the guidance and co-operation of an older geologist than himself.   It was in the fulfilment of this resolution that Sedgwick and he first became fellow-workers in the field.

Sailing packets, small boats, and post-horses combined to make a tour among the Inner Hebrides and West Highlands in those days a leisurely affair.   A geologist had many opportunities of using his hammer by the way, and Murchison seems always to have had his in his hand or in his pocket, and to have jotted down in detail what he saw. The itinerary of his journey shows that he scoured the hills and glens of Mull, peeped into every nook and cranny of Staffa, mounted to the top of Ben Nevis and recognised its curious crest of porphyry, went up to the Parallel Roads of Glen Roy, ascended the Great Glen, and then turning west through Glengarry to Glenshiel, found himself in Skye.   In that wildest and weirdest of the Western Islands he and his wife did excellent work in collecting fossils, and thereby obtaining materials for making more detailed comparison between the secondary rocks of the West of Scotland and those of England than had been attempted by Dr. Macculloch.   The actual fossil-hunting was mainly done by Mrs. Murchison, after whom one of the shells (*Ammonites Murchisoniæ*) was named by Sowerby, while her husband climbed the cliffs and trudged over the moors and crags to make out the order of succession among the secondary strata.

But the tour was not merely geological. Many a halt and detour were made to get a good view of some fine scenery, or to make yet another sketch. Friends and Highland cousins, too, were plentifully scattered along the route, so that the travellers had ample experience of the hearty hospitality of those regions. An occasional shot at grouse or deer varied the monotony of the hammering; but even when stalking, Murchison could not keep his eyes from the rocks. Amid the jottings of his sport he had facts to chronicle about the gneiss or porphyry or sandstone through which the sport had led him. This characteristic, traceable even at this early period of his life, remained prominent up to the last autumn of his life in which he was able to wield a gun or a hammer.

The summer had in great part passed before he reached that part of the eastern coast of Sutherlandshire where the scene of his special task lay; but that task proved to be eminently easy. From Dunrobin, where he was hospitably entertained, he could follow northwards and southwards a regular succession of strata, and recognised in them the equivalents of parts of the oolitic series of Yorkshire. The Brora coal, therefore, instead of forming part of the true carboniferous system, was simply a local peculiarity in the oolitic series. As in Skye, he made a collection of fossils which offered a means of satisfactory comparison with the oolitic rocks of England.

The rapidity with which this piece of work could be done left time for a prolongation of the tour northwards through Caithness, even up into the Orkney Islands, but at

length the tourists had to prepare for a southward migration
again.   Reaching Inverness, they turned eastward to Aberdeen, and thence, with Boué's *Essai* in hand, down the eastern
coast, by Peterhead, Bullers of Buchan, Arbroath, and St.
Andrews.   While in Fife they received tidings of the serious
illness of the old General at Nursted.   Abruptly closing this
protracted ramble, they took their places in the mail-coach,
and travelled without intermission into Hants.   The immediate result of this summer's work was seen in the preparation of a paper for the Geological Society.[1]

As before, the winter was passed in London, and this
became henceforth Murchison's practice.   The summer and
autumn usually found him in the country for fresh observations, with visits to old friends and a renewal of field-sports ;
but when winter began to set in, unless when abroad, he
made his way back to town to renew the socialities of life,
in which he delighted, and to elaborate his geological work
for publication.

Among the incidents of London life in the winter of
1826-27, he has preserved some notes of a hazardous descent into the Thames Tunnel, then in course of construction.   The river had burst in upon the works, and the two
Brunels were organizing means for expelling the intruder.
Considerable discussion went on in scientific circles as to
the mode of procedure, or whether it was worth proceeding
at all.   Dr. Buckland organized a party to go down and

---

[1] " On the Coal-field of Brora, in Sutherlandshire, and some other
stratified deposits in the north of Scotland" (*Trans. Geol. Soc.*, 2d series,
vol. ii. p. 293), an excellent memoir, in which the principles of William
Smith were, for the first time, applied in detail to the oolitic rocks of
Scotland, and which gave the first connected account of these rocks, with
lists of characteristic fossils.

inspect, including Charles Bonaparte (afterwards Prince of Canino) and Murchison.

" The first operation we underwent (one which I never repeated) was to go down in a diving-bell upon the cavity by which the Thames had broken in. Buckland and Feather-stonehaugh, having been the first to volunteer, came up with such red faces and such staring eyes, that I confess I felt no great inclination to follow their example, particularly as Charles Bonaparte was most anxious to avoid the dilemma, excusing himself by saying that his family was very short-necked and subject to apoplexy, etc.; but it would not do to show the white feather; I got in, and induced him to follow me.   The effect was, as I expected, most oppressive, and then on the bottom what did we see but dirty gravel and mud, from which I brought up a fragment of one of Hunt's blacking-bottles. We soon pulled the string, and were delighted to breathe the fresh air.

" The first folly was, however, quite overpowered by the next. We went down the shaft on the south bank, and got, with young Brunel, into a punt, which he was to steer into the tunnel till we reached the repairing-shield. About eleven feet of water were still in the tunnel, leaving just space enough above our heads for Brunel to stand up and claw the ceiling and sides to impel us.   As we were proceeding he called out, ' Now, gentlemen, if by accident there should be a rush of water, I shall turn the punt over and prevent you being jammed against the roof, and we shall then all be carried out and up the shaft !'   On this C. Bonaparte re-marked, ' But I cannot swim !' and, just as he had said the words, Brunel, swinging carelessly from right to left, fell overboard, and out went of course the candles, with which he was lighting up the place. Taking this for the *sauve qui peut*, fat C. B., then the very image of Napoleon at St. Helena, was about to roll out after him, when I held him fast, and, by the glimmering light from the entrance, we found young Brunel, who swam like a fish, coming up on the other side of the punt, and soon got him on board. We of course called out for an immediate retreat, for really there could not be a more foolhardy and ridiculous risk of our lives, inasmuch as it was just the moment of trial as to whether the Thames would make a further inroad or not."

As the spring months wore away, short visits to the country could be resumed, as, for example, down to Oxford, to join in one of the galloping excursions of the merry Professor of Geology, or to Lewes to make the acquaintance of Dr. Mantell, then in full medical practice, but who had found time to distinguish himself as a zealous palæontologist and collector. In the course of these short and desultory excursions, Murchison supplemented his former work in the Petersfield district, and made himself master of the full succession of the cretaceous formations.

But a much more lengthy and ambitious tour had already been planned. In the previous year, during the rambles in Arran and elsewhere in the north, he had met with many puzzling facts. Particularly had he been discomfited by the problems presented by the red sandstones of the west coast. And as we have already noted, he had determined to return to the attack, bringing with him a geologist of ampler knowledge and specially experienced in the complicated structure of the older rocks. Of all his geological friends none had won his respect and admiration so entirely as Sedgwick. Admirable as an observer, clear

and brilliant as an expositor, the Woodwardian Professor
was one of the kindliest, wittiest, merriest of companions.
While Murchison's pursuit of science was now and continued through life to be a serious earnest task, Sedgwick's
enthusiasm and earnestness, on the other hand, were quite
as great, his knowledge far greater, but he threw over
his scientific work the charm of his own bright genial
nature.   Brimful of humour and bristling with apposite
anecdote, his scientific talk was greatly more entertaining
than the ordinary conversation of most good talkers, for he
could so place a dry scientific fact as to photograph it on the
memory while at the same time he linked it with something
droll or fanciful or tender, so that it seemed ever after to
wear a kind of human significance.   No keener eye than his
ever ranged over the rocks of England, and yet while noting
each feature of their structure or scenery he delighted to carry
through his geological work an endless thread of fun and
wit.   No wonder therefore that Murchison, who, though not
himself gifted with humour, had a keen relish for it as it
came from others, should have made choice of such a companion.

But Sedgwick had already distinguished himself in the
difficult labour of unravelling the structure of some of the
older rocks of this country.   And it was in the older rocks
that the problems lay which had baffled Murchison during
his first geological raid into Scotland.   In every way the
society of the Cambridge Professor would be an advantage
to him ; it would give him at once a skilful instructor, a
generous fellow-labourer, and a buoyant companion.   His
proposal that Sedgwick should return with him to Scotland
was accepted, and the two friends, destined to achieve many

ADAM SEDGWICK, F.R.S.

From a Photograph.

an arduous and hard-won success in after years in the field together, started on their first conjoint geological tour early in July 1827.

The main object of this journey was to ascertain if possible the true relations of the red sandstones of Scotland—a subject in regard to which Murchison himself had observed many difficult or apparently contradictory facts in the previous year, and which the maps and writings of Macculloch had not fully explained.   The route chosen agreed on the whole with that previously followed by Murchison and his wife—Arran, Mull, Skye, thence through the north of Sutherlandshire to the east coast of Caithness, and then southwards by Elgin, Aberdeen, Forfarshire, Edinburgh, Dumfriesshire, Carlisle, and Newcastle, to York.

Throughout by much the greater part of the country to be traversed in the Highland tracts comparatively little had been done by geologists beyond the maps and memoirs of Macculloch, and hence there was little in the way of published description to be read before starting.   From a loose slip of paper found among Murchison's repositories, it appears that in the absence of geological memoranda he had taken to the acquisition of words and phrases in Gaelic, and had written down such as he judged would be most useful.   The reader may think this list rather an ominous one when he is told that it begins with the question in Gaelic, "Where is the public-house?" and ends off with " ooshke clay—hot water."

From this long and well-worked journey Murchison profited greatly.   Under Sedgwick's guidance he saw clearly enough now the meaning of things which had puzzled him not a little before.   For example, even at that early time,

Sedgwick had distinguished that peculiar structure in rocks
to which the name of "cleavage" is now given, and taught
his companion to recognise it.[1]   Fractures and foldings, with
other broad features of geological structure in a region of
old dislocated and altered rocks, were likewise unravelled.

But with Sedgwick in the party the tour could not possibly be all work and no play.   They threw themselves
heartily into the ways of the Highlanders, and made friends
all along the route,—ate haggises and drank whisky at one
house, danced in rough coats and hobnailed boots in another, brightened with talk and tale the drawing-room of a
third.   Much of the journey was performed on foot over
wild moor and mountain, or in a crazy boat through the
winding fjords.   Some of the expeditions too were undertaken in such storms of wind and rain as are seldom seen
anywhere in Britain out of that north-western region.   Hence
they returned to the south country, not without adventures to
boast of,—how, for example, they were nearly lost in boating
from Greinord to Ullapool, and saved, so Sedgwick said, by
his vigorous help in bailing the leaky boat with his hat,—
or how, Sedgwick wearing a plaid which he had bought from
a shepherd, they were taken by a bustling landlady for a
couple of drovers, and got but scant courtesy,—or how, to
prevent a like mistake at Forfar, Murchison insisted on

---

[1] Among the slate-quarries of Ballachulish they met with examples of
cleavage which Sedgwick pointed out on the spot to K. von Oeynhausen
and H. von Dechen, then rambling through Scotland and gathering materials for the papers on various parts of Highland geology, which they
afterwards published in Karsten's *Archiv.*   He failed to convince them
that there was any essential difference between the original stratification
of the rocks and the lines of cleavage, even though the argument lasted
long, in one of the deluges of rain so characteristic of that weeping
climate.

going first into the inn, and, to his companion's delight, was
shown into the tap-room ! from which, however, the retired
captain of dragoons discharged such a characteristically military volley of denunciation as speedily brought both landlord
and landlady with profuse apologies and a loud command of
" wax-lights for the gentlemen." Among these incidents of
travel one curious coincidence made an impression upon
Murchison's Highland susceptibilities. His mother, as we
have seen, was a Mackenzie of Fairburn, born in the ances-

Red Sandstone Mountains on the West Coast of Sutherland.

tral Tower. There had been a tradition in the district to
the effect that the lands should pass out of the hands of the
Mackenzies, and that " the sow should litter in the lady's
chamber." The old tower had now become a ruin, and the
two travellers turned aside to see it. " The Professor and
I," says Murchison, " were groping our way up the broken
stone stair-case, when we were almost knocked over by a
rush of two or three pigs that had been nestling up-stairs
in the very room in which my mother was born."

After seeing most of the red sandstone tracts of Scotland the two travellers re-entered England by Carlisle, crossed to Newcastle, and revisited some of the sporting scenes of earlier years.   One of the friends they saw was Murchison's former fox-hunting chief, Lord Darlington, who, he writes, " laughed at my new hobby which had converted me into ' an earth-stopper ! ' "—a simile worthy of a veteran Nimrod who hunted every day of the week except Sunday.

With the winter came back the usual routine of London life. The Secretaryship of the Geological Society demanded a good deal of time and labour, and the President, Dr. Fitton, kept a sharp eye on his subordinates, so much so, indeed, that an actual rupture took place between him and Murchison, which was only healed after much correspondence, and by the intervention of friends, who endeavoured to convince the President that he was too exacting, and the Secretary that he was too insubordinate. Murchison kept all the letters he received on the subject, and inscribed on the outside of the packet,—" 1827.   Some months' waste of time—Fittoniana, or disputes with my warm-hearted but peppery friend Dr. Fitton."

But besides looking after the lucubrations of other writers aspiring to geological fame, he had plenty of work this winter in extending for the Society his notes of the Scottish tour with Sedgwick.   The latter was full of work at Cambridge ; suffering, too, from weak eyes, and given to " water-drinking and dephlogisticating,"—apt, therefore, to delay what he could push aside for a time, and needing, as he said himself, an occasional nudge on the elbow.   His pen was required for the conjoint memoir as much as his hammer had been for the work in the field; but who could expect much continuous literary labour from a man who could speak of himself thus? —" Behold me now !" he says, in a letter to Murchison (28th October), " in a new character, strutting about and looking dignified, with a cap, gown, cassock, and a huge pair of bands—the terror of all academical evil-doers—in short, a perfect moral scavenger. My time has been much taken up with the petty details of my office, and in showing the lions to divers papas and mammas, who, at this time of the year, come up to the University with the rising hopes of their families. This week I have to make a Latin speech to the Senate, not one word of which is yet written. I mean to write a new syllabus of my lectures, which commence in about a week; in short, my hands are as full as they well can be. I will, however, do the best I can for our *joint-stock work.*"

The two friends had resolved to make their work in the Highlands the subject of two Memoirs for the Geological Society—one on Arran, and one on the Conglomerates of the northern and eastern counties. The former of these was at last read to the Society in January 1828, but the second was kept back by Sedgwick's delay. In a later letter he refers to a hint from Dr. Fitton to make haste, lest Murchison should forestall him, and generously speaks of their joint share in the field-work thus:—" You worked harder in many respects than I did myself, and till we reached the east coast, and indeed there also, you were my geological guide." Weeks slip away, and still no help comes from the Woodwardian Professor, who writes to his friend,—" I fear you will think me a sorry coadjutor, for all the work is left to yourself.   This is not as it ought to be, but I am at present almost a lame soldier."   Still time passes, and brings April round without the completion of Sedgwick's contribution.   On the 7th of that month he says,—"You call upon me 'for my own reputation and your peace of mind to make ready.'   I promise, if God spare my health and preserve me of sane mind, to have all in good state before the reading; but to expect that our documents should exactly tally, so that we have only to stitch them together, is to expect impossibilities.   One is making a key, and the other a lock, which never can fit till the wards are well rasped and filed.   To rasp and file will be part of my office, as well as to fit on a head and tail."   At last, on the 16th May, the conjoint paper[1] was fairly launched before the Geological Society.

Murchison had left London for the Continent before that date.   His fellow-labourer, however, sent him an account of the reception of their first conjoint work.   " Our paper," Sedgwick writes, " increased to such a size that it was ob-

[1] Among the excellent details in the paper on Arran (*Geol. Trans.*, 2d series, vol. ii.), the authors erred in identifying the various rocks with supposed English equivalents.   The structure of the island is too complex to be worked out offhand in a week or two, and some of its problems are even yet not understood.

The paper on the Old Red Sandstone of the North of Scotland likewise showed great observing skill ; but the same risk of error, from comparatively hurried examination of a few traverses, was shown in it.   The authors massed all the red sandstones of the west and east coast—an error which they committed, though knowing what Macculloch had written on the subject, and which Murchison many years later discarded. One special merit of the paper was the important announcement (confirming that made in the Brora paper), of the abundant fossil fishes found in many parts of Caithness, and the plates and descriptions given of some of the forms, which in later years were to become so well known through the writings of Hugh Miller.

viously too large to be taken in at one meeting. . . . All
went off well, and ended with the dish of Caithness fish,
which were beautifully cooked by Pentland, and much

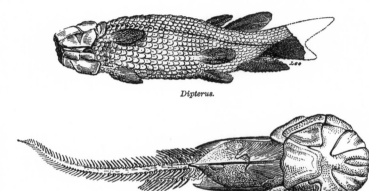

Dipterus.

Coccosteus.

Fishes of the Old Red Sandstone of Scotland.

relished by the meeting. Greenough, Buckland, Conybeare,
and all the first performers were upon the boards."

These are confessedly details of no great moment in
themselves. They seem, however, to find a fitting place
here, inasmuch as they serve to show the hearty spirit of
friendship and co-operation with which these two men
worked together in the early years of their intercourse.

CHAPTER IX.

THE three years which had now passed away since his geological hammer was first buckled on had been to Murchison a time of hard work. Even in mere physical exertion his labour had been great, and would be inadequately represented by the statement that he had trudged on foot for many hundreds of miles over rough shores and still more rugged mountains. His enthusiasm had been so thoroughly awakened that there was now no risk of desertion from the scientific ranks. He had learnt a vast deal in that short interval, and learnt it too where alone it can be truly mastered—in the field. Of the many avenues of research which the infant science of Geology was opening, he had already chosen that along which he was to rise to eminence. Whether in the south of England, among late secondary and tertiary rocks, or in the north and west of Scotland, among some of the oldest palæozoic masses, his leading aim had been to unravel the true order of arrangement of the rocks, and show their relation to each other and to those of other and better known regions. In this pursuit he felt that he could distinguish himself, and he had done so. With leisure at command and a wide field for exertion, spurred too by a real love for the work as well as by a strong desire to be prominent, his first three years of geological labour at home had been a marked success. From a mere beginner he had speedily become one of the prominent men at the Geological Society, and one of the most ardent and promising of the rising geologists of his day.

So thoroughly had geology dispossessed, at least for the time, all other occupations, that his note-books for these years contain memoranda of hardly anything else. Elaborately does he detail every section which he saw ; minutely does he describe every step and stage of each of his journeys. The main scientific results have long been given to the world, and there remains, besides the mere dry itinerary, but the scantiest residuum of personal matters to show in what other ways his thoughts and time were engaged. Among his papers occur notes of invitation—a dinner with Davy, a soirée at 'Fitton's,—or memoranda of meetings and consultations with friends of the Royal or Geological Society, and jottings enough to show that his scientific pursuits had in no way slackened his general activity and energy, or lessened his pleasure in the convivialities of society.

But having successfully essayed his strength among the rocks of his own country, it was not to be supposed that he would long refrain from making a dash at those of the Continent, where it was thought that a good deal might be done in applying the principles of classification which had been so successfully used among the Secondary rocks of England. Accordingly in the winter of 1827-8 he began to

turn his thoughts towards a foray of that kind.   The result
was that, once abroad, he found so much of novelty and
interest there as to bring him back again and again.   Hence
for the next three years the scene of his labours extended
from the Straits of Dover through central and southern
France to the shores of the Adriatic on the one hand, and
through Rhineland, Bavaria, and Austria into Hungary on
the other.

The first of these continental excursions was planned to
include the centre and south of France, the north of Italy,
and parts of Switzerland.   As usual, copious notes were
made from the various authors who had treated of the
geology of these tracts.   " I induced my wife," he writes,
" to accompany me as well as my associate, Charles Lyell.
We were off in April, and on the 26th of that month were
at work in the field with Constant Prevost, following his
subdivisions of the Paris basin.   The theoretical views of
Prevost made a deep impression on Lyell, who, as far as I
can judge, imbibed some of his best ideas of the operation
[*sic*] of land and fresh water alternations with marine deposits from the persevering and ingenious Frenchman."

At Paris they met also Cuvier, Brongniart, Deshayes,
Élie de Beaumont, Desmarest, Dufrénoy, and other scientific
men of mark, and made further notes for the summer's work.
By the beginning of June they found themselves among the
wonderful extinct volcanic cones of Auvergne.   This singularly interesting region had been admirably described shortly
before, both with pen and pencil, by Mr. Poulett Scrope,
whose memoir they carried with them.   They were fortunate,
moreover, in having an introduction to Count Montlosier,
one of the noblesse of Auvergne, who, while taking part in the

political struggles of his country, had devoted himself also
to the study of the volcanic rocks of that district, which he
had described with great spirit and accuracy. Amid the
troubles of the time he had lost all his property, "except a
portion of mountain which was too ungrateful a soil to find
another purchaser." Retiring to this retreat in his old age he
had built himself a cottage in an extinct crater. "The traveller in approaching the door of the philosopher of Randane
had to wade through scoriæ and ashes;" but beyond these
obstacles he found a hospitable roof and a host whose "lofty
and vigorous presence accorded well with his frank and
chivalrous demeanour." [1]   A hearty welcome awaited our
three tourists. Their coming had been anticipated by the
old Count, from whom on reaching Clermont they found
awaiting them a note of invitation and welcome (still
extant) couched in that tone of mingled dignity, courtesy,
and cordiality which seems now one of the lost arts. "He
was charmed to see us," records Murchison, "and to go over
all his old volcanic subjects, and instruct us on every feature
around his residence, except on the post day when his
papers and letters came. Then he flew to them, excusing
himself with the old French politesse, ' Pardonnez, Messieurs et Madame; mais c'est ma vie.' " [2]

The three gentlemen, on foot or on horseback, and Mrs.
Murchison on a stout pony of the Count's, explored together
the cones of cinders and *cheires* of lava. Even to one who
is familiar with volcanoes the first sight of these marvel-

[1] Whewell, *Proc. Geol. Soc.*, iii. 70.
[2] The Count Montlosier "died in 1837, at the age of eighty-three, on
his way to Paris to take his seat in the Chamber of Peers, of which he
was a member." See a brief sketch of him by Dr. Whewell, in the
address referred to in the preceding note.

lously fresh cones and craters and lava-rivers fills the mind
with astonishment. He wanders perhaps up a narrow and
picturesque valley feathered with birch and broom down the
sides, and gaily green with meadow and orchard along the
bottom.   Suddenly he comes upon the rough black lava,
usurping the channel of the stream, and still bare and
bristling, as if it had only yesterday stiffened into rest.   And
then climbing further by the edge of the lava-torrent, he
comes at last in sight of the marvel of the region—the chain
of Puys—cones of volcanic materials still so perfect that he is
tempted to watch if steam or smoke cannot still be seen rising
from their tops.   But when, crossing the lava stream, he
mounts the steep sides of one of these old volcanoes, he finds it
cold and silent.   There beneath him lies the crater—a deep
hole sunk into the summit of the hill, no longer breathing
out volcanic heat and fumes, but carpeted even to the
bottom with turf, and fragrant with many a wild-flower.
And from these depths, whence in old times came the snort-
ing and bellowing of the volcano, there rises now on the
breeze only the tinkle of the cattle-bells or the hum of the
bee.

These are the youngest of the volcanoes of Central France,
but all round them lie fragments of older and yet older
eruptions, pointing to a long protracted volcanic period—so
long, indeed, that the rivers of the district had been able to
cut out in the older lavas deep and wide valleys, down which
some of the later lavas flowed.   Beyond measure instructive,
therefore, is such a country to the geologist, inasmuch as it
places before him admirable illustrations of the action both
of subterranean and external forces.

Amid such scenes as these, our travellers spent some six

weeks, riding, climbing, driving, and filling note-book and
sketch-book with memoranda of rocks and scenery. These
rambles bore fruit during the succeeding winter in papers
which were read before the Geological Society.[1]

Turning eastward, the travellers journeyed leisurely down
the valley of the Rhone, looking at rocks and antiquities by
the way, until they reached Montpelier, and thence passed
on by Nismes to Aix, in Provence.[2] After quitting Toulon,
an incident occurred to mar the good spirits and hinder the
work of the party. Murchison caught a malaria fever, and
became rapidly delirious. He soon recovered, however, and,
except a temporary loss of strength, suffered no evil effects,
escaping more fortunately than his wife had done, for the
symptoms of the fever she was seized with at Rome used to
return upon her at intervals all through life. To recruit him
a halt of nearly three weeks was made at Nice, where the
invalid soon regained his former activity, scouring the district all round the town under the guidance of Risso the
conchologist, who led him over the fossiliferous deposits.

While recruiting his health at Nice, Murchison sent an
account of the tour to the Woodwardian Professor, from
which a few sentences may be quoted. In Central France
" we left various things undone, consoling ourselves that
such a case was to be worked out by Sedgwick next year.
And here let me, by way of parenthesis, invoke the philosophical spirit of inquiry which prevails at Cambridge, and
urge *you*, who are really almost our only mathematical

[1] " On the Excavation of Valleys, as illustrated by the Volcanic Rocks
of Central France." By Charles Lyell and Roderick I. Murchison.—
Proc. Geol. Soc., i. 89. See also p. 140.
[2] See *Proc. Geol. Soc.*, i. 150, where their conjoint paper on this tract
is given.

champion, not to let another year elapse without endeavouring to add to the stock of your British geology some of
the continental materials. Pray do it before you marry and
settle for life ; pray even do it before you bring forth that
long-expected second volume on the Geology of England
and Wales ;[1] your comparisons will then have a strength
and freshness which will quite electrify us." " We met
with splendid cases of basalt and trap, rivalling in antiquity of aspect our northern acquaintances," " splendid
proofs of the extraordinary amount of excavation in the
valleys," two thousand feet or more of fresh-water strata,
with apparently " everything which characterizes even the
older secondaries"—" red sandstones," " grits, shales," " an
excellent cornstone, and beneath this *lymneœ* and *planorbes ;*"
little " coal-fields—true chips of the old coal-block." " In
dust and insufferable heat, which have never quitted us
since, we descended the Rhone." " The only cool place we
could find was Buckland's hyæna cave at Lunel. Our
journey across to Aix en Provence was most interesting,
and that place offered so much that we halted a week, our
work being now reduced to four or five hours in the morning, from four to nine, and a little in the evening. We
hope to show you twenty or thirty species of *insects !!* from
the gypsum quarries there. In this city of idleness we
have been pent up during ten days, not daring to travel
into Italy with these heats : it has not rained one drop here
for eight months."

After making a number of excursions together in the
Vicentin, Mr. Lyell having finally resolved to abandon law
and devote himself wholly to geology, turned off southwards

---

[1] Conybeare and Phillips' *Outlines* being considered the *first* volume.

to pursue his inquiries among the tertiary rocks, while the other two travellers struck eastwards to Venice, and thence into the Alps.   At Bassano, Murchison collected materials for a paper on the tertiary and secondary rocks of the Tyrolese Alps, which was read to the Geological Society in the following spring.   Ascending by Botzen, he examined the now well-known earth-pillars—tall pyramids of stony clay, each with a stone or big boulder on its summit, and conjectured their materials to have been accumulated by " powerful torrents coincident with the elevation of the chain."   At that time the former extension of the glaciers of the Alps had not yet been realized by geologists.   Hence not at Botzen only, but up the valley of the Inn, and in other parts of the mountains traversed in this tour, Murchison, following the prevalent notions of the time, looked upon all the masses of " drift," with travelled blocks, as the results of powerful deluges or *débâcles*, which swept down the valleys or over the hills.

Having recently supplied the Geological Society with what Sedgwick called " a dish of fossil fish" from the old red sandstone of Caithness, he took the opportunity of turning aside to collect another meal of the same materials from the bituminous schists of Seefeld—a little mountain village of the Tyrol, where some of the rocks were so impregnated with animal matter, from the abundance of fish remains imbedded in them, that for generations the villagers had been in the habit of roasting fragments of the stone, out of which they obtained oil for their lamps and cart-wheels. This little episode was turned to account in the following winter, and bore fruit in a paper upon these dark schists and their fish, read to the Geological Society.

A leisurely journey, with many halts by the way to allow of the use of hammer and sketch-book, brought the travellers through the picturesque tract between the valley of the Inn and the Lake of Constance, and thence once more into Switzerland. But this time it was not fine scenery, nor even a field for feats of pedestrianism, which formed the chief attractions of the country. At every resting-place an attempt was made to ascertain the nature and sequence of the rocks, and as much time and labour were now given to hunt up an old quarry as in former days would have been gladly given to find out a half-hidden specimen of an old master. Reaching Stein, Murchison set at once about exploring the quarries of Oeningen, famous for having formerly yielded the skeleton which Scheuchzer gravely described as " Homo diluvii testis ;" but which more recent science has shown to be not human, but salamandrine. " To my joy I learnt," he writes, " that in the last two years the quarries had been re-opened, and that a very remarkable new quadruped had been recently exhumed. This splendid fossil had fallen into the hands of a doctor and a silversmith of the little town, and was in the house of the former, where I inspected it, and counted twenty-three vertebræ. On the whole it was like a dog, fox, or wolf. I resolved at once to acquire it, provided, on my return to Paris, M. Cuvier should pronounce upon its value, the sum asked being £30. It was however, essential that I should have a drawing, and therefore my wife stole out with her pattens across the muddy street early next morning, before the doctor was up, and induced the servant girl to let her in to sketch the beast. The moment Cuvier saw the drawing he said it was in all probability a fox. Of course an old fox-hunter like me could not resist the *bonne bouche* of finding the first fossil fox, and, writing back from Paris, I acquired the animal, which I gave to the British Museum,[1] and which Owen has since turned into the ' dog of the marsh,'—more nearly related to the civet-cat than any other living animal." [2]

Journeying by Basle, Strasbourg, the Vosges mountains, and thence through France, with many a stop and detour to visit geological sections or the contents of museums, the travellers did not reach England until the end of October. They had thus been six months abroad. During that time Murchison seems to have done his best not to let a single day pass without adding to his stock of geological knowledge. With an enthusiasm which must have made him a somewhat troublesome companion, he spared no bodily fatigue in pursuit of his inquiries, throwing himself as heartily into questions regarding the order of succession among the rocks of each town or valley he visited, as if the place had been his home. The work of these six months was reduced to form in two memoirs, which he himself prepared in the succeeding winter for the Geological Society, and in three conjoint papers written in concert with Mr. Lyell. But the results are to be measured not so much by these published records of them as by their influence in finally clenching his geological bent, and fixing him in that stratigraphical groove in which he had made his first essay in the south of England, and in which, with but short and not altogether successful deviations, he was to pursue his geological career to the end.

[1] The counterpart slate he gave to the Geological Society.

[2] Professor Owen named this unique specimen *Galecynus Oeningensis*, and regarded it as belonging to " an extinct genus intermediate between *canis* and *viverra*."—See *Quart. Journ. Geol. Soc.*, iii. (1847), p. 60 ; and *Palæontology*, 2d edit., p. 412.

The winter of 1828-9 was spent as usual in London. The preparation of the five memoirs just referred to, as well as the business of the Secretaryship of the Geological Society, kept Murchison's hands full enough of work.[1] " M. Valenciennes," he notes, "was in London this winter and helped me to describe the fossil fish of Seefeld, and I was gathering knowledge from Stokes, Broderip, Wollaston, Buckland, Greenough, Lindley, Curtis the entomologist, König, Webster, and Mantell." He found time, however, to do a little field-work now and then, for in visiting friends in the country he came no longer simply as a sportsman. Some of the notes of invitation of these years occur among his papers, and show that his new zeal for stones furnished many a point for a quiet joke at his expense, where the writers, while referring half deprecatingly to the use which they could wish to see him make of his gun, are at pains to assure him that he need not want opportunities of wielding his hammer.

With spring and the prospect of fresh work in the field plans were vigorously sketched for a new campaign. Again an attack on the structure of the Alps was decided upon, but this time it was not to be single-handed. Professor Sedgwick had agreed to share in the toil and glory of the warfare, having determined to quit for a time his books at Cambridge and his vacation rambles at home, and trust himself with his hypochondria to the rough fare of unfrequented routes abroad. It was again Murchison's task to collect all the information obtainable from papers or friends as to the geology of the tracts to be visited.

[1] Among his note-books there is one with detailed notes of a series of lectures on the structure of birds, which he attended during the spring of 1829.

In June the two travellers set out together, and travelling rapidly by Bonn, the volcanic tract of the Laacher See, Coblenz, and Cassel, halted at Göttingen to geologize. There they chanced by a curious coincidence to stumble upon their two Prussian friends, von Oeynhausen and von Dechen, with whom they had held the fierce argumentation in a deluge of rain at Glencoe. " I was just about to sally out," Murchison writes to his wife, " when little Oeynhausen popped his nose into the room where S. and self were dressing. In an instant we were in each other's arms, and I can assure you that he kissed me on each cheek at least a score of times. And the Professor did not come off with a short allowance. Think of our good luck ! He with his *nouvelle mariée*, mother-in-law, and Dechen with his *sposa* are here. The vivacious little Prussian discovered me by *the name upon my hammer*, as it hung out of the old stone-bag in the carriage-yard." Again, he records that at Göttingen " Our hero (Sedgwick specially rejoiced in him) was old Professor Blumenbach, then eighty-six years of age, on whom we called. He told us loads of amusing anecdotes. Among his numerous skulls he showed me one of a Highlander sent to him by Sir George Mackenzie, and he denied that my countrymen had higher cheek-bones than other people. We afterwards attended his lecture of the day on insects, and were astonished at his versatile powers, his extraordinary action, his fine deep voice, and impressive countenance. Whether he rolled out hard words with all the rapidity of a youth, or thumped his desk with all the vivacity of a youth, or suddenly paused abruptly to explain with a broad slow ' aber, aber,' before he finished by some reservations, I looked at him as the most original of God's

works I had ever seen.   As I had presented him in the morning with some of my fossil insects from Aix, he launched out
in illustration of these flies and bugs which had lived ' vor
Menschen,' and then carried his pupils off to the British
Museum and our gigantic Scarabæus in granite.   Drinking
tea with him in the evening, Blumenbach equally astonished
us by his extensive reading and wonderful memory, whether
he adverted to metaphysics and Bishop Berkeley, to Scottish
history and scenery and Walter Scott, or the vitrified forts
and Sir George Mackenzie."[1]

Turning northward the two travellers made their way
through the Harz Mountains and thence by way of Halle to
Berlin.   At that early time the older palæozoic rocks were
all classed together under the uncouth title of " grauwacke,"
and among Murchison's notes reference is made to the "interminable grauwacke," which deprived so much of the
journey of geological interest.   Strange that before many
years passed away it was among such rocks that he earned
his chief title to scientific fame, and that they offered attraction enough to lead him hundreds of miles from home, and
to keep him busy over mountain and valley for months
together !   This very region of the Harz, as we shall find,
furnished, only ten years later, abundant interest and plenty
of hard work for the two fellow-labourers among these same
grauwacke masses.   In the meanwhile, however, these rocks
seem to have had somewhat of a depressing effect upon
Murchison's spirits, so that the wit and sparkle of the Professor were never more welcome.

The halt at Halle brought them in contact with a real

[1] A brief biographical sketch of this remarkable man will be found in
vol. iii. of the *Proceedings of the Geological Society*, p. 533.

living specimen of a staunch Wernerian in the person of
Professor Germar, who expounded the geology of the country
after the system of his master, no doubt to the infinite
delectation of the Cambridge Professor, who must have
looked upon the old theorist as an interesting relic of a
species of geologist that was gradually becoming extinct.
But they succeeded in picking up a few scraps of information regarding some of the regions included in their programme of travel, and their visit to Berlin was similarly
successful.

Southward the journey lay by Dresden through Bohemia
to Vienna and the confines of Hungary, and thence by the
caves of Adelsberg to Trieste, "a hot hole, although it has some
luxuries in it—good ice and water-melons that would make
any man ill except Sedgwick." From that point, which was
the limit of their journey, the travellers bent their steps
homeward again through the Carinthian Alps, the Tyrol, and
the Salzkammergut, striking westward into Switzerland by
the Lake of Constance, and descending the Rhine to Strasbourg, whence they found their way across France, so as to
reach England once more in the end of October.

Some of the pleasantest days of this tour were those in
which the travellers enjoyed the society of that remarkable
man, the Archduke John, among his mountain retreats in
Carinthia. "Our chief object in coming to Gastein," Murchison writes, "was to wait upon the most scientific Prince
in Europe, the Archduke John, and he received us with cordiality and frankness. We dined at the rural *table-d'hôte*,
at which the landlord presided, carved, and could boast
with pride that his ancestors had kept the inn for 350 years.
At this board, besides the Archduke, we had imperial minis-

ters and generals, Prussian nobles, as well as professors and
geologists.	After dinner we set out to ascend, in a *char à
banc*, with the Archduke and his chamberlain, to the upper
cascades at Naasfeld.	We passed the village of Böckstein,
where the gold ore is washed, and thence viewed the snowy
range of the Ankogel, to the summit of which the Archduke had ascended, viz., 10,000 feet high, and seven hours'
good walk above the highest châlet.	We reached the upper
fall at sunset, and were then in the region of summer-
châlets, and surrounded by snowy peaks and glaciers, the
boundary between Carinthia and the Salzburg region.

"The Archduke was a capital cicerone, and talked
familiarly with every one we met.	One of these was a
rough Carinthian packman, whose broad lingo amused us,
and reminded me of Goldsmith's line—

'Or onward where the rude Carinthian boor;'

though I do not think that Oliver, for the sake of rhyme,
had any right to add—

'Against the houseless stranger shuts his door.'

Nor would the Archduke allow that they were a bad set of
fellows, though very inferior to his Styrians and Tyrolese.
All the miners were 'hail-fellow' with the Prince—*i.e.* with
perfectly good manners, but with no *mauvaise honte*.

"On our homeward trip on foot we had a *petit souper* of
fresh trout, which the Archduke had ordered for us in the
village of Böckstein, and in approaching the cabaret several
peasant girls ran out with their little nosegays, and to kiss
his hand; whilst he of course put the flowers into his broad-
brimmed Styrian hat.	As we walked down the valley in a
fine starlight night we had much enlivening chat, and we
soon perceived how honest a liberal the Prince was.	He

laughed at all the old stiffness and prejudice of the Austrian court, to the dress of which his Styrian jacket, black leather shorts, and long green worsted stockings presented a marked contrast. He is a first-rate chamois-hunter, and kills about forty bucks annually. . . . He talked with delight of everything in his dear Styria,—the clean inns, honest inn-keepers, and pretty waiting-maids. He specially abused all men-waiters, who had found their way to Grätz, and whom he stigmatized as ' des hommes de deux maitres '—*i.e.* as waiters and ' agens de la haute police.'

"Next morning we were at the door of the Archduke by appointment at 7. It was opened by a bluff Styrian jäger, who beckoned us into the curate's small sitting-room, then the only residence of his Imperial Highness, whom we found on his knees, his hob-nailed boots taken off, and busily at work laying òut on the floor the Austrian trigonometrical map of the surrounding Alps for our inspection. Showing us all the passes, he gave us many good instructions."

The scientific fruits of this expedition have long been before the world. They were given to the Geological Society in four successive papers. during the succeeding winter and spring. Such rapid work among the broken and contorted rocks of a complicated geological region could not but contain many errors. Yet it must remain as a striking example of keen and quick observation, and of often happy, though not always accurate, generalization. In addition to their researches on the structure of the Austrian Alps, the travellers were struck by two classes of facts which could not but arrest the notice of men whose geological types had hitherto been mainly English. In the first place, they found thick beds of good black coal, masses of millstone, oolite, and

other hard rocks, to be not older than some of the soft
tertiary sands and clays at home. Well might Murchison
write—" Away went all our old notions of mineral terms
applied to geological formations as any indications of their
age." In the next place, they were again and again arrested,
and as it were appalled, by the formidable ravines and chasms
which bear witness to the enormous yearly waste of the Alps.
At one part of the course of the Fella they noticed that a
single night of heavy rain had buried the roadway under a
vast pile of rubbish swept down from the mountain-sides.
" As there are countless such torrents rushing down into the
Tagliamento and its tributaries, which is one of the six chief
rivers that flow into the Adriatic between Trieste and Venice,
we can well imagine how that sea must be encroached
upon, and at what a rate the sides and gorges of the Alps are
wearing away."

In another respect the tour had not been without its
fruits. It brought the two English geologists into direct
personal relation with the geologists of Germany, from whom
they received much kind attention and assistance. A groundwork was thus laid for much pleasant and friendly intercourse in later years. In passing through France too they
formed or renewed acquaintance with several brethren of the
hammer in that country, notably with M. Élie de Beaumont, whom they met at Boulogne, and from whom, then in
the early enthusiasm of his pentagonal theory, they received
details regarding the order in which he supposed the mountains of the globe had been elevated—details, however,
which their own work among the Alps would hardly
support.

The winter months of 1829-30 were spent in London,

where the duties of the Secretaryship of the Geological Society, the preparation of his memoirs on the recent Continental tour, and the ordinary but increasing social exigencies of his position, kept Murchison's hands fuller than ever of work, though he still found now and then an opportunity of escaping to the country to visit a friend and have a few days' shooting. Indeed, it would seem from a letter addressed to him in March that the old fox-hunting Adam was not yet wholly cast out of him.

Nevertheless when summer had brought back sunshine and flowers to the Alpine valleys, he determined to revisit them.

On the appearance of the abstracts of their papers on the Austrian and Bavarian Alps in the Proceedings of the Geological Society, the views which Sedgwick and Murchison had put forth were combated in British and foreign journals, notably by Dr. Ami Boué. Before the publication of their completed memoirs, the two fellow-labourers saw clearly that to meet the objections which had been urged, it would be necessary for one or both of them to revisit a few of their sections, and to examine some of the new localities which had been cited as adverse to their views. Murchison gladly undertook this congenial task. Accordingly, early in June he started with his wife, primarily for the purpose of clearing up these difficulties, but also to see a little more of German scenery and society as well as German geology and geologists.

The tour lasted until the beginning of October, and embraced, besides the old ground, some parts of Europe which he had not yet seen since he had taken to scientific pursuits. Crossing to Ostend, and proceeding by Antwerp to Brussels

and Namur, where he "was enraptured with Omalius
d'Halloy;" Liége, where young Dumont, just beginning his
career, lent the traveller his services; Cologne and Bonn—
Murchison sped up the Rhine without any halt for geological
exploration. At that time he still "despised the old slaty
rocks," though before another year was over he was to begin
the forging of that chain which kept him to them for the
rest of his life. " I was then keen on one scent only, viz.,
greensands, chalk, and tertiary," and it was to study these
rocks yet more fully that he had again set out for the eastern
Alps.

Instead, however, of striking at once into the mountains,
the travellers made a detour through Bavaria, passing by
Aschaffenburg, Bamberg, Bayreuth, and Ratisbon, to Vienna.
Every museum on the way was examined, and notes were
made of its contents in so far as they might throw light upon
the secondary rocks of the Alps and surrounding regions.
Every local geologist too seems to have been ferreted out
and pressed into service. At Bamberg, by good chance, a
name of more than local celebrity caught Murchison's eye
in the visitors' book at the inn.   " I instantly rushed to
the museum," he writes, " where I introduced myself to the
great geologist to whom Humboldt and all Germany bowed
—Leopold von Buch. We had at once a most interesting
colloquy on dolomitization and many of the recent discoveries.
The little vivacious man was then quite *en tête* with his
monograph of Ammonites. Though turned of sixty, he had
only of late begun to study organic remains, and at once he
was endeavouring to generalize and group these animals by
their sutures. I perceived at once how with all his great
qualities, he was irascible if any contemporary criticised him,

and he was then in a particular rage about Buckland's
having omitted to state that the bear-caverns of Muggendorf
and Gailenreuth were in pure dolomite ! He had just undergone a severe penance, owing to his obstinacy in never
taking a guide.   He was lost in a forest on a stormy night,
and passed the hours of darkness under a tree, with no protec-
tion but an umbrella which he then always carried.   As he
got old, however, he threw even that aside, and braved wet
and cold in a plain black suit, and without any change of
garments."[1]

At Vienna, besides museums, picture-galleries, and geologists, Murchison saw a good deal of "distinguished society,"
for which to the end he had a special fondness.   He renewed
his acquaintance with the Archduke John, dined with Lord
Cowley, ambassador at the Austrian Court, and had an opportunity of holding converse with Metternich.   He has preserved a record of part of the conversation at the ambassador's table.   The talk had drifted into geology, and a lady
present—the same who had been the heroine in the incident
at dinner in Messina (*ante*, p. 53)—asked across the table a
question about science and the Mosaic record.   "I naturally
had some difficulty in getting out of the dilemma, when
Metternich, taking up the cudgels, gave them to my surprise
a capital lecture, and quite to the purpose.   On going into
the drawing-room after dinner, and on sitting down on the
sofa to converse with the great diplomatist who had over-
thrown Napoleon, I soon learnt how and where he obtained

[1] It was not until further experience of Continental geology and
geologists that Murchison conceived that great respect for Leopold von
Buch which he used often to express in his later years, adding at the
same time a cordial recognition of what he conceived to be his own obligations to the influence of the German geologist.

his geological knowledge. ' You will not believe me (said he) when I tell you that I love science more than politics. In my early youth I took honours in scientific studies, and intended to give up my life to such pursuits, and become a Docteur-ès-Arts et Sciences.   But the French Revolution startled all the old Austrian families, and my father insisted on it that as I had a name to sustain, I must, for the good of my country and the honour of my family, betake myself to public life.   So I was sent as an attaché to the embassy at Paris.   There, in the intervals of business, and when not occupied in the study of the doings and character of Napoleon, I was always an attendant at Cuvier's lectures. The words of that great master have never been forgotten, and hence my repetition of them, when I supported you at table, and showed to my diplomatic friends the great *usefulness* of your science, for that is the only mode of approaching them.'

"In his conversation he showed that he had read and thought much on this subject, and particularly on the application of geology to the development of the mineral wealth of Austria.   He endeavoured to make me believe that he was all in favour of a scientific meeting in Vienna next year, following those of Hamburg, etc., which had already taken place.   He expressed his ardent hope that the people would become more scientific, and hoped that I would publish some work upon their country, and stir them up a little.

"When I told the Archduke John afterwards of this conversation of Metternich's, he said it was all *fudge*, and merely intended to blind me !"

Breaking away at last from these attractions in the

capital, Murchison betook himself to the serious work which had been the main object of the journey. He had written to Sedgwick that in order to prove their points he would, if possible, " riddle these Alps in all directions"—a resolution which he now proceeded to put in practice. Accompanied by Professor Paul Partsch, an active geologist of Vienna, he made several minor excursions in the neighbourhood, and then, striking through the Leitha Gebirge as far as Grätz, turned back westwards into the Alps.[1] The wonderful little tertiary basins enclosed among the older rocks of Carinthia, and sometimes furnishing thick masses of lignite, first detained him. But the real hard work lay among the mountains of the Salzkammergut and Styria, the object being to clear up the relations of the supposed tertiary strata of Gosau and the structure of the secondary rocks of that part of Austria. In the state of the science at the time, it was no wonder that Murchison, though making out some new points in the structure of the mountains, still missed the meaning of the curious and puzzling assemblage of fossils at Gosau. Several weeks of very hard work were spent in those regions, with the result of confirming some main parts of the conjoint survey of the previous year, and of showing the need to modify others. From Ischl, in the midst of the rambling, he wrote to Sedgwick :—" O, what would I give that our sketch of the Alps was not out ! I could make it so much more perfect in details and sections. . . . All these points necessarily involve important alterations in our sections, which I

---

[1] Some excellent observations were made during this time on the age of the older rocks of Carinthia. They have been recently referred to by M. de Koninck in his " Recherches sur les Animaux fossiles," 2de Partie, 1873, p. 2 (*Sur les Fossiles Carbonifères de Bleiberg*).

hope have not been begun.   After a great deal of hard work
I have relieved my mind from a world of anxiety, and am
now resting and thankful, and taking a vapour salt bath or
two, enjoying right worshipful high Vienna society, who
are all stewing themselves in salt here.   I am at same
time working out the details of the upper beds (upper grits
and marlstones of the Alpen-kalk), which by a charming
accident I have got within half a mile here."

About three weeks later the same correspondent received
a further detailed narrative of geological exploits in a letter
dated from Sonthofen, and beginning thus :—" Here I am,
sticking to my scent like a true fox-hound.   Since I wrote
to you from Ischl I have done some marvellous good work.
I made out a fine range of the Gosau beds near that place.
. . . At Hallein I found V. Lill all anxiety to see me. . . .
The moment I twigged certain secondary black fossils like
lias (in his den near the river), and ascertained that the
section was not above a six hours' excursion, the post-wagen
was ordered, and off we travelled. . . . I soon made a most
clear and instructive section, with lias shells and sufficient
fossils to make out the case. . . . How I did pant and fag
on the north side of Untersberg, for which I had glorious
hot weather.   I made four parallel transverse sections.   I
think I have the whole thing now most clear : it is certainly a capital key."

"I set out with a heavy heart to cross 120 miles of
Bavarian pebbles, and exactly 100 back to Augsburg, in
order that I, Rod. I. M., should heal my pricking conscience
and that of my dear ' heilige freund' Adam Sedgwick *in re*
' Sonthofen.' . . . I flatter myself I get to understand the
valley, but with devilish ado and many perplexities—nay,

more than I ever encountered in my geological career. My
throw off occasioned a hearty malediction upon Herren
Sedgwick and Murchison, who as they drove up to Sonthofen last year passed through a certain archway leading
into that valley, with a rock close to them which they never
hammered. This I found to be true genuine old greensand.
. . . But when I came to go along the south flanks of the
Grinten, and ascend to the iron mines, all my precognosced
friends seemed to be sent topsy-turvy. What inversions
and contortions ! . . . I left no gorge nor any mountain
peak unexamined where I thought examination necessary."

Quitting at last these puzzling rocks on the flanks of
the Alps, he turned homeward by Munich, Nüremberg,
Gotha, and Göttingen. At Nüremberg he notes in his
journal " a change of scene : fossils and rocks were forgotten
for a day or two." Curiously enough, however, in the next
sentence he writes—" A picture of Luther reminded me of
Buckland in his jolliest moments, while the pensive and
reflective Melanchthon is well represented in England by
Henry Warburton." In Gotha he " passed an evening with
the most remarkable man of the place, Von Hoff, whose
works on physical geography and geology proved afterwards
of such good service to Lyell."

On the 1st October Mr. and Mrs. Murchison set sail
from Rotterdam for London. And thus ended one of the
pleasantest of the continental rambles which they had yet
undertaken. They had accomplished the definite object
which had given point and aim to the journey, and had
besides seen much new country and made many new
acquaintances. The tour was, moreover, the last of this
early foreign series. The next nine years were to be em-

ployed at home in laying the foundations of that Silurian
system by which the name of Murchison will be chiefly
remembered in the history of geology.

Before we turn to that point of the narrative, the work
of the winter of 1830-31 remains to be very briefly noticed.
During the preceding three years Murchison had filled many
note-books with innumerable memoranda of sections, fossil
collections, excerpts from published descriptions and verbal
information, all bearing upon the geology of the secondary
rocks of Germany.   The long and elaborate memoir of
Sedgwick and himself on the eastern Alps, still in the press,
would, when published, contain all the main points of their
work; but many details remained which it seemed desirable
to publish, especially in so far as they might bear upon
English geology.   To carry out this idea, and verify some
parts of the larger memoir, he went to Paris to compare a
collection of fossils from Germany, and partly, as he con-
fessed, "to frequent the society of scientific friends."   With
Alexander von Humboldt, who happened to be there at the
time, he made acquaintance, and got from him much information regarding some of the geological aspects of the great
geographer's travels.

How the foreign materials were produced at the Geological Society may be partly gathered from the subjoined
letter to his friend Sir Philip Egerton (28th January 1831):—

"I am quite vexed that I should fire off all my Alpine
crackers without your hearing the report of one.   I finish
on Wednesday next, when the whole of the meeting-room
will be hung with sectional tapestry of the manufacture of
Lonsdale[1] and Co., magnified from my smaller designs.   If,

[1] The worthy Curator of the Society's collections.

therefore, you have any intention of being in town for the meeting of Parliament, being Friday, perhaps you can accelerate your movements (particularly as it freezes hard), and be with us ; otherwise you will miss a golden opportunity of learning how much deposit took place between the periods of our English chalk and London clay, and throughout such extensive regions that I verily believe our case in Western Europe will prove to be the exception and not the rule. Besides this, I will warm you with basaltic eruptions which, though they only show the tips of their noses, have heaved up mountains of gneiss and granite against the greensand series, setting it, and the tertiary strata above it, all on end.

"I was out of town for a fortnight, shooting at Charles Lefevre's, and at Up Park about the Christmas time, since when I have been working like a slave, previous to quitting office—not with disgrace, however, as my friends are going to vote me into the President's chair, in which case I shall request you to be one of my councillors—a post well befitting so grave a senator. Our anniversary, when all the jollification and election take place, is the 18th February— so you may bow to the Queen in the morning, and to me at night."

CHAPTER X.

FOR five years the Secretary of the Geological Society had worked energetically for the Society's behoof, catering for papers, arranging the reading and publication of them, and preparing, either alone or in conjunction with the Woodwardian Professor of Cambridge or Mr. Lyell, some able memoirs on structural geology. He had earned a claim to the Society's gratitude, which was acknowledged this winter (February 1831) by his election to the dignity of President. The chair had been previously filled by Sedgwick, who, on quitting it, concluded his address with these words :—" Mine has been indeed but an interrupted service ; but I resign it to one of whose powers you have had long experience, who can give them to you undivided, and whose hands are in no respect less ready than my own."

The office is held for two years. How it was filled by Murchison will be told in the next chapter. We have now arrived at the great turning-point of his scientific life, and must look at it with some care, that its bearings may be clearly seen not only on his own career, but on the history of geology.

Up to this time, his work in the field had lain almost

wholly among Secondary rocks, whether in this country or abroad, insomuch that, as we have seen, the rocks of older date seemed to him to wear a dry, forbidding aspect, no matter where they might present themselves. But before the close of the first session of his Presidency at the Geological Society he had determined to look these old rocks steadily in the face, and see what after all might be their meaning and history. Every year brought fresh and often apparently contradictory facts to light about them. They evidently deserved to be studied, and would probably reward any adventurous spirit who chose resolutely to grapple with their problems. Murchison, at the instigation of Buckland and other friends, made up his mind to try.

The labours which have now to be traced as they went on year by year, have a far wider interest than merely their relation to the life and work of the man by whom they were conducted. They unquestionably established a notable epoch in the progress of geology. They added a new chapter to geological history. They have been of infinite service in helping the interpretation of what are called the palæozoic rocks in every quarter of the world. To gain an adequate notion of what they were and how they came to acquire the importance now justly ascribed to them, we may cast our eyes first of all, and very rapidly, over the knowledge, or rather the ignorance, which existed in this part of geology before the date of Murchison's researches.

Over the centre and south of England the great series of rocks now embraced under the term "Secondary" have undergone comparatively little disturbance from those subterranean movements which have in other regions heaved up these same rocks into some of the loftiest mountain-chains

upon the surface of the globe. They lie one upon the other
with almost the regularity of the shelves in a library.
Their story, therefore, when once the key to decipher it had
been given, was not difficult to read. The genius of William
Smith had supplied that key, and thus the investigation of
the Secondary rocks had made such enormous strides during
the previous fifteen or twenty years, that it seemed as if
little more could be done in that branch of geology, save to
elaborate details. Starting from the types of the undisturbed formations of England, men endeavoured by their
means to reduce into order the complicated structure of such
regions as the Alps. Among those who successfully essayed
such a task, Murchison had taken an honourable place.

But down below these Secondary rocks, and underneath
the Carboniferous and Old Red Sandstone deposits, the succession of which had been made out by William Smith, there
lay others, so hardened, squeezed, and broken as seemingly
to defy all attempts to classify them by the same minute and
detailed method. Such rocks stretched over most of Wales, of
Devon and Cornwall, of the Lake Country, and of the uplands
of the south of Scotland. They covered wide spaces on the
Continent, as for instance in Scandinavia, Rhineland, and
Bohemia. It was known that they must be enormously thick.
From year to year an increasing number of the remains of
corals, crinoids, shells, and other organisms was reported
from them. Evidently, therefore, they did not all date from
a time anterior to the introduction of life upon the earth.

Many were the names given to this vast and hetero-
geneous series of rocks. That proposed by Werner had met
with the widest acceptance, viz., Transition—a name which
implied the theory that these rocks had been formed at a

period of the world's history transitional between a time when rocks were laid down all over the globe by chemical precipitation from a hot ocean, and a time when conditions more like those at present in force permitted of the existence of living creatures upon the earth.

Another appellation which had been very generally applied to these old rocks was "grauwacke"—an uncouth word originally used by the Harz miners for a special kind of rock in

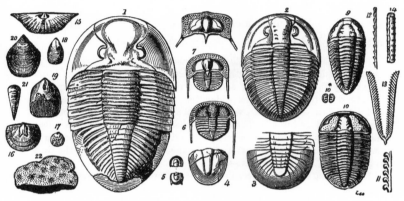

CHARACTERISTIC FOSSILS FROM THE GRAUWACKE (LLANDEILO FLAGS).

1-10. Trilobites. 1. Asaphus tyrannus. 2. Ogygia Buchii. 3. O. Portlockii. 4. Stygina Murchisoniæ. 5. Agnostus Maccoyii. 6. Trinucleus fimbriatus. 7. T. Lloydii. 8. T. concentricus. 9. Calymene brevicapitata. 10. C. duplicata. 10*. Beyrichia complicata. 11. Graptolithus Beckii. , 12. G. tenuis. 13. Didymograpsus Murchisonii. 14. Diplograpsus teretiusculus. 15. Orthis alata. 16. O. striatula. 17. Siphonotreta micula. 18. Lingula attenuata. 19. L. granulata. 20. L. Ramsayi. 21. Theca reversa. 22. Monticulipora favulosa.

the Transition series, and gradually adopted as a convenient name for a great part of the most ancient stratified masses. But though often used as if it signified a particular division of geological time, grauwacke was really the name of a particular rock, and hence wherever that rock occurred, the name might be legitimately given to it, without reference to respective age, or under the mistaken impression that all grauwacke was of the same general geological date.

Under such vaguely applied names, rocks of vastly different ages and characters were incongruously grouped together. Hence they presented so many contradictions and difficulties that geologists on the whole avoided them as much as possible. Murchison only reflected the common dislike of them when he hurried through the Rhine provinces to get away from what he called the " interminable grauwacke." Writers of text-books were sorely puzzled how to marshal the few discordant facts which were already known on the subject. Fanciful theory and mere trim mineralogical distinctions often supplied the place of geological knowledge.[1]

[1] No better illustration could be obtained of the state of this part of geological science at the time than the fact that the *Principles of Geology* of Lyell, while devoting about 300 pages to the Tertiary deposits, dismissed all fossiliferous rocks older than those above the coal-measures in twelve lines.—(*Principles*, vol. iii., published in the spring of 1833, and dedicated to Murchison.) The account there given of these rocks does not pretend to be more than a reference, but it may be quoted here as a curious commentary on the state of ignorance which prevailed at the time regarding the Palæozoic rocks:—

" 6. *Carboniferous Group, comprising the coal measures, the mountain limestone, the old red sandstone, the transition limestone, the coarse slates and slaty sandstones called graywacke by some writers, and other associated rocks.*

" The mountain and transition limestones of the English geologists contain many of the same species of shells in common, and we shall therefore refer them for the present to the same great period ; and consequently the coal, which alternates in some districts with mountain limestone, and the old red sandstone, which intervenes between the mountain and transition limestones, will be considered as belonging to the same period. The coal-bearing strata are characterized by several hundred species of plants, which serve very distinctly to mark the vegetation of part of this era. Some of the rocks, termed graywacke in Germany, are connected by their fossils with the mountain limestone."

The third edition of a popular English geological text-book—Bakewell's *Introduction to Geology*—appeared in the year 1828, and contained the following table of the rocks now referred to :—

" TRANSITION CLASS (Conformable).

" 1. Slate, including flinty slate and other varieties.

2. Greywacke and greywacke slate, passing into old red sandstone.

3. Transition limestone. Mountain limestone."

In the third edition of the excellent *Geological Manual* of the late Sir

When we consider the extremely perplexing character of the geology of many of the districts where these old rocks occur, we cannot wonder that they should have continued to be a stumbling-block in the progress of the science. The key furnished by William Smith for the secondary rocks might not have been found for many years later, if these strata had lain less regularly in England than they do. To men who came fresh from such undisturbed deposits to the contorted, fractured, and hardened older rocks, it must have seemed well-nigh a hopeless task to reduce the apparent chaos to order. Professor Sedgwick,

Henry De la Beche, all the fossiliferous rocks under the old red sandstone are thrown into the "Grauwacke Group," which is described as "a large stratified mass of arenaceous and slaty rocks, intermingled with patches of limestone, which are often continuous for considerable distances. The arenaceous and slate-beds, considered generally, bear evident marks of mechanical origin, but that of the included limestones may be more questionable." The fossiliferous character of the group is insisted on, and 126 genera and 547 species of fossils are enumerated from the grauwacke rocks of this and foreign countries. When, however, we look into these fossil lists, we find that a large number of species belong to rocks which are now placed on the horizon of the old red sandstone or Devonian system, and that others have been inserted which should have been placed on the still higher horizon of the carboniferous limestone. The confusion of the lists is only a faithful reflex of the utter confusion in which the stratigraphy of the rocks themselves still lay.

Even as late as the year 1832, after Sedgwick had published his views as to the structure of the transition rocks of the Cumberland district, and after Murchison had made known the distinct order of succession in the upper portions of these rocks around the Welsh border, the able and well-informed Conybeare could report to the British Association but a meagre statement of the scanty knowledge then obtained on this part of British geology, and is found gravely discussing the "need of a term less barbarous than grauwacke-slate, which would conveniently denominate the characteristic rock of this era. Might not clasmoschist (from the Greek $\kappa\lambda\alpha\sigma\mu\alpha$) be conveniently adopted? It would afford a term well contrasted to mica-schist, the characteristic rock of the primitive group."—(*Brit. Assoc. Reports*, vol. i. p. 382.

On the Continent the ignorance was quite as dense as here, although, appearing under the guise of hard names and neatly arranged tables, it

indeed, nine years before the time at which we are arrived, viz., as far back as 1822, had begun to grapple with the rocks of his Cumbrian mountains, and, in spite of their broken and contorted character, was slowly unravelling their structure.   But no amount of labour or skill in that region could possibly connect the history of the Transition rocks with that of the younger strata by which they are covered; for a great gap occurs there in the geological record, which is thus rendered as imperfect as a historical narrative would be if several important chapters were torn out of it and destroyed.   A similar hiatus had been so frequently observed elsewhere that the notion had become general that the so-called " Transition " rocks belonged to a totally different and distinct order of things, and that they had been fractured and upheaved before any of the Secondary formations were laid down upon them.

Any attempts which had been made to subdivide the Transition series, and to connect those of one country with those of another, had been based hitherto wholly on the

might have passed for exact knowledge.   Thus the *Elémens de Géologie* of J. d'Omalius d'Halloy, offered the subjoined table to its readers as showing the most advanced views in the year 1831 :—

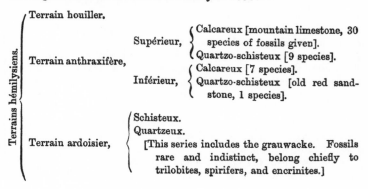

Terrains hémilysiens.	Terrain houiller.		
	Terrain anthraxifère,	Supérieur,	Calcareux [mountain limestone, 30 species of fossils given].
			Quartzo-schisteux [9 species].
		Inférieur,	Calcareux [7 species].
			Quartzo-schisteux [old red sandstone, 1 species].
	Terrain ardoisier,		Schisteux.
			Quartzeux.
			[This series includes the grauwacke. Fossils rare and indistinct, belong chiefly to trilobites, spirifers, and encrinites.]

mineralogical characters of the strata. But these characters, as is now well known, afford no sufficient test of geological age and position, the grauwackes and shales of one age being often in that respect undistinguishable from those of another. Besides, even when used in reference to one continuous series of rocks, though often most convenient and useful, they are liable to constant and rapid changes. They could not, therefore, be safely relied upon for a sound and generally applicable classification, such as had been established by means of fossil evidence among the overlying formations.

And yet the transition rocks were far from being destitute of fossils.[1] These were to be had sometimes in great abundance. They seemed to be in the main of peculiar species, not found in the overlying strata. Hence it was evident that before any use could be made of the fossils in the way of grouping the rocks into divisions, the very order of succession among these rocks had first to be settled. But no one who had hitherto addressed himself to this task had been able to establish as a basis for palæontological work any broad and serviceable divisions among the old grauwacke, or to connect it satisfactorily with the formations

[1] Their fossiliferous character had been noted by Werner. In England fossils had been found by William Smith and Mr. Phillips in the uppermost Transition rocks of Westmoreland. These specimens were shown to Sedgwick in 1822, and slightly described by him in his paper on Craven in 1827. The fossiliferous character of some parts of the Transition series of Shropshire and Wales was likewise well known, though no one seems to have set about determining what the fossils were, and how far they agreed with or differed from those of the overlying formations. "Practically," to quote from some notes obligingly furnished by Professor Phillips, " before the summer of 1831 the whole field of the ancient rocks and fossils of Wales was unexplored ; but then arose two men— *par nobile*, of all men fitted for the purpose—Sedgwick and Murchison— and simultaneously set to work to cultivate what had been left a desert."

which succeeded it in time.   So broken indeed and altered
was it that if any one had proposed to apply to this puzzling
old transition or grauwacke series the same tests by which
the secondary and tertiary deposits had been brought into
such clear and intelligible order, he would have raised a
smile among his geological friends.   Murchison knew of
course no more about these ancient formations than his
neighbours, but he now resolved with his characteristic
energy and enthusiasm to see what he could make of them.

At the end of the session of the Geological Society he
started from Bryanston Square with his " wife and maid,
two good grey nags and a little carriage, saddles being
strapped behind for occasional equestrian use."   Some
preliminary skirmishing took place among the secondary
and tertiary rocks by the way, for he could not resist the
sight of a quarry or pit, being resolved to miss nothing on
the road.   The route lay by Oxford, where his old friend
and preceptor Buckland received him, and led him over
some of the ground where he had formerly received his
earliest lessons in field geology.   But it was not merely to
renew old acquaintance that a halt was made at Oxford.
" I took notes from Dr. Buckland," he writes, " of all that
he knew of the slaty rocks, or grauwacke as it was then
called, which succeeded to the Old Red Sandstone, and the
relations of which I was determined to begin to unravel;
and I recollect that he then told me that he thought I
would find a good illustration of the succession or passage
on the banks of the Wye east of Builth."

This laudable custom of collecting all available information, published or unpublished, regarding any piece of
geology, before himself attacking it, has already been fre-

quently apparent in the preceding narrative. It came forward prominently enough at the commencement of this new and momentous enterprise. He had already made notes in London, while Dr. Buckland furnished him with new and valuable suggestions. Quitting Oxford, he journeyed westward to visit the Rev. W. D. Conybeare, a name honourable in the history of geology as that of one of the joint authors of the *Geology of England and Wales.* From this kind and experienced friend he notes that he obtained " some good advice." Other local observers, who, though not aspiring to be called geologists, had been in the habit of looking at the rocks and fossils of their neighbourhood, gave him invaluable assistance. Among these helpers may be mentioned Dr. Dugard of Shrewsbury, Mr. Anstice of Madely, Dr. Lloyd of Ludlow, Mr. Davies of Llandovery, and above all the Rev. T. T. Lewis of Aymestry. From the first these friends enlisted readily in his service, and some of them continued their unremitting toil and kindness for years. To Mr. Lewis especially he was indebted for much of his knowledge of the rocks and fossils of the upper Silurian series, for that gentleman had made out the arrangement of the rocks in his district, and recognised their characteristic fossils before Murchison had begun to study the subject.

On first taking the field this year Murchison had spent some time in a desultory series of visits to country friends and rambles after Secondary strata. His companion during a portion of the time was Mr. Phillips, who has given the following notes of the journey :—" In the cool spring-time of 1831 we met by appointment at Staneford, and explored together the district of Collyweston and Ketton. It was

a pleasant walk along the high grounds overlooking the
Willand; cigars contending with endless discussions on the
rocks around us, and on their relationships to Alpine limestones which had begun to be recognised. We made careful measures of the slaty and sandy beds full of shells which
here overlie the ironstone and the lias, and intended to give
a joint memoir as to their position and numerous fossil contents. Collyweston has been again and again visited by me,
but not I think by Murchison, who in that year had his
attention drawn to a larger field of work, and began to dream
of Siluria."

The dream was soon to become a reality. For, crossing at
last to Swansea, Murchison struck northwards into the hills
beyond the coal-field, and there began to invade the Transition rocks of South Wales. These hills consist of the
Carboniferous Limestone rising out from under the Coal
measures and resting upon thick masses of Old Red Sandstone, so that when one crosses the high ground and
descends into the lower regions towards the north, one comes
upon lower and lower strata cropping up from beneath the
Old Red Sandstone, and spreading for many a league over
the undulating country to right and left and in front. It
was near the town of Llandeilo that Murchison first broke
into these older rocks with the purpose of making them dis-
close their true place and order in the geological series.

"Travelling from Brecon to Builth by the Herefordshire
road, the gorge in which the Wye flows first developed
what I had not till then seen. Low terrace-shaped ridges
of grey rock dipping slightly to the south-east appeared on
the opposite bank of the Wye, and seemed to rise out quite
conformably from beneath the Old Red of Herefordshire.

Boating across the river at Cavansham Ferry, I rushed up
to these ridges, and to my inexpressible joy found them
replete with transition fossils, afterwards identified with
those at Ludlow.   Here then was a key, and if I could
only follow this out on the strike of the beds to the north-
east the case would be good."

To and fro through the Welsh and border counties he
worked his way as the rocks led him northwards over hill
and valley into the plains of Cheshire.   The expedition was

Vale of the Towy, from near Llandeilo. (Sketched by Mrs. Murchison.)

far more successful than he had dreamed it could be, for, by
a happy accident, he had stumbled upon some of the few
natural sections where the order of the upper parts of the
transition rocks in Britain can be readily perceived, and
where their strata can be traced passing up into the overlying formations.   No one could better appreciate the value
of this "find" than the fortunate geologist himself.   "For
a first survey," he writes, " I had got the upper grauwacke,
so called, into my hands, for I had seen it in several situa-
tions far from each other all along the South Welsh frontier,

and in Shropshire and Herefordshire, rising out gradually and conformably from beneath the lowest member of the old red sandstone. Moreover, I had ascertained that its different beds were characterized by peculiar fossils. I had, therefore, quite enough on hand to enable me to appear at the first meeting of the British Association, which I had promised to join at York in October, with a good broad announcement of a new step in British geology."

His notes, however, show that he did not rush at once from the grauwacke to the York assembly, but journeyed so leisurely as to pay many visits to old north-country friends, and to fill up long pages of jottings by the way on the geology of the region between the hills of Wales and the sea-coast of Durham. At last, the same " pair of greys" which had carried the two travellers from London all through the Welsh border, and the midland and northern counties, deposited Mr. and Mrs. Murchison at the hospitable gates of Bishopthorpe, where they remained as guests of the Archbishop during the first meeting of the British Association. Of that memorable meeting, so important an event in the history of science in this country, Murchison has preserved the following recollections :—

" FIRST MEETING OF THE BRITISH ASSOCIATION AT YORK,
27th September to 3d October 1831.

" This first gathering of men of science to give a more systematic direction to their researches, to gather funds for carrying out analyses and inquiries, to gain strength and influence by union, and to make their voice tell in all those public affairs in which science ought to tell, came about in this wise :—Assemblies of ' Naturforscher' had been for two years or more in existence in Germany, having begun in Hamburg. Thereon Sir D. Brewster wrote an article in the *Edinburgh Philosophical Journal* suggesting that such a meeting should be tried in Britain. On this the Rev. William Vernon (afterwards Vernon Harcourt), the third son of the Archbishop of York, and a Prebendary of York, not only made the real beginning by proposing that we should meet at York, but by engaging his father to act as a Patron, and by inducing Earl Fitzwilliam to be the President, he gave at once a *locus standi* and respectability to the project. But he did much more ; for he elaborated a constitution of that which he considered might become a Parliament of Science, such as Bacon had imagined, and was thus our lawgiver.

" The project thus elaborated having been transmitted to me in London in the spring of 1831, when I was President of the Geological Society; I at once eagerly supported it. Nay, more, I wrote and lithographed an appeal to all my scientific friends, particularly the geologists, urging them to join this new Association. But notwithstanding my energy, the scheme was for the most part *pooh-poohed*, and, among my own associates, I only induced Mr. Greenough, Dr. Daubeny, Sir Philip Egerton, and Mr. Yates, to follow suit. John Phillips of York, the nephew of William Smith, and the Curator of the York Museum, had very much to do in the origin of this concern, for he co-operated warmly with William Vernon, and, when we got together at York, was the secretary and *factotum*. He had previously corresponded with me in London, and stimulated me with a ready-made prospectus. I may say that it was the cheerful and engaging manners of young Phillips that went far in cementing us ; and even then he gave signs of the eminence to which he afterwards arose in the numerous years in which he was the most efficient assistant-general-secretary of the body, until when, as the distinguished Reader of Geology in the University of Oxford, he presided over the British Association at Birmingham.

" When, however, we were congregated from all parts, the feebleness of the body scientific was too apparent. From London we had no strong men of other branches of science, and I was but a young President of the geologists; from Cambridge no one, but apologies from Whewell, Sedgwick,[1] and others; from Oxford we had Daubeny only, with apologies from Buckland and others. On the other hand, we had the Provost of Trinity College, Dublin, Dr. Lloyd, Dr. Dalton, from Manchester, and Sir David Brewster from Edinburgh. Thus there was just a nucleus which, if well managed, might roll on to be a large ball. And admirably was it conducted by William Vernon, for, after opening the meeting in an earnest, solemn manner, the good Lord Fitz-william handed over the whole control to Harcourt and left us.

" On my own part I had plenty of matter wherewith to keep my geological section alive, as, besides those I have mentioned, we had a tower of strength in old William Smith, the Father of English Geology, and then resident at Scar-borough; James Forbes, Tom Allan the mineralogist, and

---

[1] " Sedgwick indeed sent his apology through me, in a letter from Llan-fyllin. It was his *début* among the North Welsh rocks. ' Cracking the rocks of Carnarvonshire for three weeks, and getting fond of the sport,' he writes, ' I should be a traitor to quit my post now that I am keeping watch among the mountains. It would be very delightful to mingle among the philosophers and commence deipnosophist, but it would be very bad philosophy in the long-run. You may tell Mr. W. Vernon that keeping away is a great act of self-denial on my part, and that I am in fact doing their work by staying away.' "

Johnston the chemist from Edinburgh, to say nothing of Harry Witham of Lartington (now an author on fossil flora), and others, including William Hutton of Newcastle-on-Tyne, then strong upon his ' whin-sill.' After all, however, we were but a meagre squad to represent British science, and I never felt humbler in my life than when Harcourt, in his opening address, referred to me as representing London !

" Indeed, William Conybeare, afterwards Dean of Llandaff, had quizzed us unmercifully, as well as W. Broderip and Stokes, and other men of science. The first of these had said, that if a central part of England were chosen for the meeting, and the science of London and the south were to be weighed against the science of the North, the meeting ought to be held in the Zoological Gardens of the Regent's Park ! It required, therefore, no little pluck to fight up against all this opposition, and all I can claim credit for is, that I was a hearty supporter of the scheme—*coûte que coûte*.[1]

" This first gathering was in short much like what takes place at small Continental meetings—we had no regular sections, but worked on harmoniously with our small force *in cumulo*. The excellent Archbishop was of great social use, and gave a dignity to the proceedings, whilst Lord Morpeth, then the young member for Yorkshire, incited us by speeches as to our future. It was then and there resolved that we were ever to be *Provincials*. Old Dalton insisted on

---

[1] As an illustration of the kind of taunts amid which the British Association was born, the following sentence may be quoted from a letter written by J. G. Lockhart, editor of the *Quarterly Review*, to Murchison just before the meeting :—" I presume you are going to the colt-show at York. Don't make a fool of yourself among these twaddlers, who must, in such strength of re-union (considering what happens in all their minor associations), be enough to disturb the temper, if not brains, of the σοφωτατοι, of which number is of course the P. G. S. L."

this—saying that we should lose all the object of diffusing knowledge if we ever met in the Metropolis.

" With all our efforts, however, we might never have succeeded had not my dear friend Dr. Daubeny boldly suggested (and he had no authority whatever) that we should hold our second meeting in the University of Oxford ! ! It was that second meeting which consolidated us, and enabled us to take up a proper position. Then it was that, *seeing the thing was going to succeed,* the men of science of the metropolis and those of the universities joined us."

A letter written by Murchison from York, towards the close of the meeting, to Dr. Whewell, gives a glimpse of the enthusiasm with which some of the fellow-labourers worked for the Association :—

" Before I entered into the ' British Association ' which the meeting at York has given rise to, I was very desirous of weighing the men who were eventually to carry us through. I was really very mainly induced to join it in consequence of your letter to William Vernon, and I was quite decided in so doing when I saw the calibre of the men he had assembled, and the promises of support from those who could not attend. . . . Brewster really astonished every one with the brilliancy of his new lights, old Dalton, ' atomic Dalton,' reading his own memoirs, and replying with straightforward pertinacity to every objection in the highly instructive conversations which followed each paper. . . . I had no memoir ready myself, and did not intend to rob the Geological Society of anything intended for them, but I found that a poor and hard-working druggist of Preston,[1]

[1] Mr. W. Gilbertson (see *Brit. Assoc. Rep.,* 1831-2, p. 82). The shells referred to are in the museum of the Geological Society.

Lancashire, who had made some years ago a very important observation on the existence of shells of existing species in the gravels and marls of Lancashire at 300 feet above the sea, and at distances of fifteen and twenty miles from the sea, was present. I took the opportunity of turning lecturer, and having visited those parts this summer, I brought out my little druggist with all the éclat he merited. This is another practical exemplification of the good arising from such a reunion. The Archbishop had all the party on one of the days, and it would have gratified the liberality of Cambridge to have seen old Quaker Dalton on his Grace's right hand. Pray act cordially with us, and if Adam [Sedgwick], my great master, and yourself will only go along with us, the third meeting will unquestionably be at Cambridge. Rely on it, the thing *must* progress, all the good men and true here present are resolved to make it do so."

Fresh from the field, Murchison had not had time to prepare any important paper to inaugurate the birth of the new Association. But besides bringing forward the finder of the Lancashire shells, he took the opportunity of showing the general nature and tendency of his recent work, by hanging up the maps which he had used that summer in his tour, and on which he had coloured " the Transition Rocks, the Old Red Sandstone, and Carboniferous Limestone," etc., an exhibition of interest to geologists, since it was the first which gave promise that the uncertainty of the true relations of the Transition rocks to the later formations was now at length to be dispelled.

At the close of the meeting the "pair of greys," which had done such good service already, were again in requisition

---

[1] *British Association Reports,* vol. i. p. 91.

to transport the travellers to the east coast.   There, at Scarborough and its neighbourhood, Murchison once more availed himself of the ever ready co-operation of the illustrious " Father of English Geology," and renewed his acquaintance with the rocks of that interesting coast line.   In a letter written at that time to Mr. Phillips, he reports the first germ of a proposal which in its completed form did honour to the men who made it, and to the Government which carried it into execution.   It was one of the earliest of a long series of kind-hearted acts to meritorious but often poor men of science—acts which, if they had not Murchison for their originator, never failed to find in him an active and influential supporter.   We can picture him among these Yorkshire cliffs, with the kindly old man, who, though he had done more for geology than any man then living, was spending the remainder of his days in humble quiet at Scarborough.   And those who knew Murchison will recognise how well fitted this sight was to touch him into active and considerate benevolence.

" I have had a nice field-day with your uncle at Hackness.   What is your opinion, your real opinion, as to what *I* or my friends *could really do for him* (*i.e.* for his *benefit*) ?   It would never do to bring him to town without something sure and good was offered.   If we could persuade the Government to give him a little salary to be geological colourer of the Ordnance Maps published—do you think I ought to suggest this ?   I ask this as a preliminary : it would certainly be of national importance to have these well done, and lodged in the Tower and Geological Society."

This proposal, as we shall see, was not a mere matter of form or of transient good-will.   But before any further

WILLIAM SMITH, LL.D.
From a Portrait by Foureau.

action could be taken, the writer of it had to find his way back to London. This he did in the usual circuitous way which a geologist chooses, travelling through Lincolnshire and Norfolk in search of geological sections. While at Norwich he received from his friend Whewell a pressing invitation to visit Cambridge on the homeward journey, and as part of the attraction, was told that " You will find Sedgwick full to the teeth with Welsh porphyry and grau-wacke, and shall hear the legend of his fight with some of the old spirits of the mountains, who made a great resistance to the process of being geologized—an operation for which there is no name I believe in any of the dialects of the Gaelic ; but you know best." It was a curious coincidence that the two brother geologists should each independently have broken ground in Wales in the same year.[1] Sedgwick unfortunately had begun the attack in a region of great complication, Murchison, on the other hand, had been lucky enough to begin in one of comparatively easy comprehension. This accidental difference indirectly led the way to that sad estrangement which remains to be told in future chapters.

This had been in many ways a busy and important year

---

[1] The following extract from a letter of Sedgwick's to Murchison, 20th October 1831, gives us an interesting glimpse into the state of the work when the eager Woodwardian Professor began it in North Wales :— " The weather became so bad that I was driven out of Carnarvonshire before I had quite finished my work ; but, God willing, I hope to be in North Wales next year before the expiration of the first week in May, and with five months before me, I shall perhaps be able to see my way through the greater part of the Principality. If I live to finish the sur- vey, I shall have terminated my seventh or eighth summer devoted exclu- sively to the details of the old crusty rocks of the primary system. What a horrible fraction of a geological life sacrificed to the most toilsome and irksome investigations belonging to our science ! When I finished Cumber- land I hoped some one else would have done North Wales, but I have been disappointed. *N'importe.* I am now in for it, and must go on."

in Murchison's career.   He writes of it thus :—" In sum-
ming up what I saw and what I realized in the summer of
1831, or in about four months of travelling, I may say that
it was the most fruitful year of my life, for in it I laid the
foundation of my Silurian System.   I was then thirty-nine
years old, and few could excel me in bodily and mental
activity.   'Omnia vincit labor' was my motto then, and I
have always stuck to it since."

CHAPTER XI.

WHEN once more back at his post in London, it was one
of Murchison's first cares to prosecute further the scheme
for doing honour to William Smith. How his plan pros-
pered is best told in his own words, as written at the time to
Mr. Phillips :—" You know all my heart's desire for our
good old father in geology. I propounded the same (as
expressed to you) to the Council of the Geological Society
at our first meeting in November, and I only waited for the
gathering of the men of office to sound Lord Morpeth on the
feasibility of my plan, and, if approved of by him, then to
throw in a strong memorial to the Government. Judge of
my delight then, when I found that Lord Morpeth had
anticipated my wishes, and had already written to Lord
Lansdowne, arguing Smith's merits, and asking for a small
pension. This application I was asked to second, which I
have done by letter a few days ago to Lord Lansdowne; but in doing this I have deviated so far from the
original request, as to point out to Lord L. that Mr.
Smith was *still* capable of doing the State *good* service. I

went into an *exposé* of the whole thing, and proposed the
creation of a new appointment, with some such title as
' Geological Colourer of the Ordnance Maps '—thereby
meeting all the objections and criticisms of the Humists
which might be directed against sinecure places or pensions,
but which could not hold good with respect to an office so
connected with the development of the mineral wealth of
the country as that which I have suggested. We shall see
what the Lords will do, and in the meantime we had better
say nothing of it to Smith."

They had not a long time to wait, for the Government
granted the venerable geologist a pension of £100 a year
without stipulating that he should colour any Ordnance
maps.[1]

His position as President of the Geological Society
required Murchison's presence in London during winter,
even if his enthusiasm for the science and devotion to the
Society had not been amply enough to insure his attendance.
He might well be proud of the choice which the Society had
made. Thirty years later a friend of his referred to him at
one of the anniversaries of the Society as a man " born to fill
chairs." During that busy interval he certainly merited
the description. But in 1831 he sat for the first time as a
leader among his scientific brethren, in the chair which had
been held by such men as Greenough, Macculloch, Buckland,
and Sedgwick.

It was always a great object with Murchison, as President, to get what he called " a good meeting," that is, one
with interesting papers attracting a full audience, and calling

[1] For particulars of this incident, see Professor Phillips' interesting
Life of William Smith, p. 117.

out a brisk discussion. In his letters to friends in the country at this time the doings at the Society usually figure largely. For instance, writing to Dr. Whewell on the 17th November he says :—" We had a *capital* meeting last night. 1st, A memoir on the gigantic Plesio of Scarborough. 2d, Old Montlosier on Vesuvius, which drew out a long and lucid explanation from Necker de Saussure; Lyell, Buckland, Fitton, Greenough, De la Beche, and others being orators. Buckland filled up all the parts wanting in the Plesio, and perfected a monster for those who in a snowy November night were disposed to nightmare."

Certainly in those days the meetings of the Geological Society must have been among the most enjoyable gatherings in London. There was a freshness about the young science, and men still fought about broad principles, intelligible and interesting to most listeners. The inevitable days of subdivision and detail had not yet come. "Why not contrive to be here on Wednesday?" writes the President to one of his Council. "Dine with us[1] at the Crown and Anchor, and attend our meeting, where we shall have the rare union of old Adam of Cam, Buckland, Conybeare, etc." Rare union indeed! The only paper read at the meeting was by Sedgwick—one of those luminous efforts which by a few broad lines served to convey, even to non-scientific hearers, a vivid notion of the geology of a wide region, or of a great geological formation. Embalmed in the Society's printed publications, the paper, as we read it now, bears about as much resemblance to what it must have been to those who heard it, as the dried leaves in a herbarium do to the plant which tossed its blossoms in

---

[1] *i.e.* The Geological Club, to be immediately referred to.

the mountain wind. The words are there, but the fire and humour with which they rang through that dingy room in Somerset House have passed away.

In several of the learned Societies, and among them the Geological, there had sprung up what were called "clubs;" these were gatherings of the more prominent members to dine and talk, and thereafter to adjourn to the evening meeting of the Society. Besides promoting good-fellowship among the members, they gave opportunities for much pleasant scientific gossip, and, what was one of their most important functions, they kept up a strong nucleus for the Society's ordinary meetings, to which, after a comfortable dinner, the club adjourned in a body. Murchison, at this time, and to the end of his life, took a leading share in the business, gustatory and other, of the Geological Club, which was founded in 1824. In one of his letters he urges a friend to allow himself to be proposed for this club, "which we endeavour to keep select, where you will always meet some of the choicest spirits, and where you really always pick up much geology in a quiet way."

To preside at such meetings must have been one of the pleasantest duties a scientific man of that day could perform. But over and above his ordinary work for the Society, the position of President brought with it an accession of other multifarious duties and engagements. Professor Phillips recalls how "men of science who visited London in 1831 were sure to be courteously met by the President of the Geological Society, then residing in Bryanston Place, profuse in hospitality and full of hearty zeal and kindly sympathy for his brethren of the hammer, of whatever country, which never left him."

But besides these pleasant ways of using his influence, there sometimes arose others where he was called on to take part in less amicable intercourse. Thus, one of the most notable incidents in the scientific doings in London in this year was a keen battle over the Chair of the Royal Society—a battle into which Murchison seems to have thrown himself with all the ardour of his military youth. He gives the following account of it :—" On the retirement of Mr. Davies Gilbert from the chair, a certain clique in the Society got up the notion that the Duke of Sussex would be the best person we could fix upon. As soon as the plot got wind, the indignation of all the real men of science knew no bounds, and they resolved to start Herschel as an opponent to the Royal Duke. We subscribed our names to a public protest; about eighty or ninety names were appended, including those of nearly all the notable and working men in science. It was resolved to beat us, and the greatest influence was used politically, royally, and socially to bring up voters for the so-called royal cause. I became an active canvasser for Herschel.

" At that time the Royal Society was very differently composed from what it now is. Any wealthy or well-known person, any M.P. or bank director, or East Indian nabob who wished to have F.R.S. added to his name, was sure to obtain admittance, by canvassing and by being elected at any ordinary meeting. The consequence was that over all that class of our body the Royal and Government influence of the day was overpowering, and even Lord Holland, though the gout was on him, was carried up into our meeting room, where he had never been before, to vote for his royal friend!

" I stood at the top of the stairs at Somerset House, doing my best to catch a vote as any friend ascended. We were beaten by 119 to 111. Many persons who had seen our public declaration had felt so sure we should be victors that they did not come up from the country. But so it was.

" The election over, the good Duke found himself in a dilemma. He wisely saw that he could not govern the Society if he could not make up a better Council than he came in with in 1830. He therefore resolved to choose his advisers from among those who had most stoutly opposed him, and who in fact mainly represented the science of the body. Overtures were made to myself, and I deemed it to be my duty to accept office under a Prince who could act so liberally and kindly towards his opponents."

The ground on which this latter step was justified may best be gathered from the following letter :—

" *November* 14, 1831.

" MY DEAR WHEWELL,—Oh for a quiet life ! I thought like a simpleton that reform and cholera were enough to glut one with horrors, and my poor and only consolation was that I might absorb myself in science, and so fossilize my mind and frame as to allow all those shafts to pass by innocuous. Our campaign geological opened well with an excellent memoir by Dr. Christie. . . . The point of irritation is nothing in our own good Society, but consists in the formation of a new Council for the Royal, on which they have placed my name as well as your own. I will begin with the end, which is, that after much conflicting reasoning with myself I have agreed to be on the Council, and I need not add, that my determination was mainly influenced by finding we were to have a strong battery, in which I could never disgrace myself in performing the part of a simple bombardier. . . . You know as well as all my friends with what zeal I opposed His Highness's election, but I am not of that school who would cherish a rancorous and perpetual hostility. . . . I have got over all my other scruples, and intend to go along with things as they are, and not to fight against the stream and old time by joining B. and his cold and comfortless crew. In taking this step I feel that I shall be liable to the kind innuendos of some of my *ultra* friends, but my most intimate friend Lyell, who is the only man in my confidence on the point, completely approves of my conduct."

" In this way," to return to the narrative, " the second Council of the Duke of Sussex's administration was formed. With his *bonhomie*, his ready access at all times when in health, and his earnest desire to do what was best in the interests of science, we who had been his opponents became his best friends in the sequel. There was also this advantage in having him for our chief, that all scientific rivalry was at an end.

" As an active member of the Athenæum Club (of which I was one of the original 300), I had a finger in most things which were stirring among men of letters, art, and science. It was for these men that the club was set up, Davy, Croker, and Reginald Heber being its real founders and earliest trustees. I must say that it was then a truly sociable and agreeable society. Little home dinners of twelve or fourteen were frequent, Heber or Davy often presiding, particularly the former."

The Presidency of the Geological Society was employed

by Murchison in a very characteristic way, wherein he continued to distinguish himself up to the end of his life.   He
made it the ground for gathering at his house, in a more
public and official form than one could do in a private
capacity, assemblies in which scientific men mingled freely
with representatives from that non-scientific society of rank
and fashion to which he had always been so strongly
attached.   To these gatherings Mrs. Murchison lent her
cordial help, giving them a charm which added much to
their popularity.   We shall see in the records of later years
how marked this social habit became, and what an important bearing it had upon the position of science in the society
of his day.

One of the tasks of the President during his two years'
tenure of office, is to prepare an address for the Anniversary
of the Society in February.   It had been customary to
devote that address to a general survey of the progress of
geology at home and abroad during the previous year—a
labour which in the infancy of the science was not very
arduous, and had proved to be in the highest degree useful.
Murchison had now to undertake this task, perilous though
it might be for one who only eight short years before was
known merely for a keen sportsman, as ignorant of science
and as indifferent to its attractions as any other of the north-
country squires.   Nevertheless he accepted the duty and
discharged it well.   His address, indeed, lacks the vigour,
originality, and eloquence of his predecessor Sedgwick.[1]   He
contents himself with a sober outline of the work which had
been done by the Society, and other labourers in this country

---

[1] Yet it had the advantage of revision by Sedgwick, one or two
effective touches being due to his pen.

and abroad. But he shows in every page the enthusiasm with which he now pursued geology, and gives us pleasant glimpses of the zeal and good-fellowship which marked the first generation of the members of the Geological Society. His concluding sentence runs thus :—" Permit me to offer you my heart-felt wishes for the continuance of your triumphant career, and to assure you that I consider myself truly ennobled in having been placed, for a time, at the head of a brotherhood united for purposes so great, and knit together by such lofty and enduring sympathies." [1]

As illustrative of the progress of Geology in Britain at the time, it may be mentioned that in this address the President had an opportunity of noticing Sedgwick's labours (already referred to) among the rocks of Cumberland and Westmoreland, Trimmer's discovery of marine shells on Moel Tryfane in Wales, the appearance of Lindley and Hutton's *Fossil Flora*, of the second volume of Lyell's *Principles of Geology*, and of Macculloch's *System of Geology*, the establishment of the British Association, and the great increase in number and vigour of local scientific Societies. To the thoughtful student of the history of science there is something eminently suggestive in this conjunction of the works of Lyell and Macculloch. The pages of the former writer glowed with all the fervour of the newer school of geology, which sprang out of the teachings of Hutton and William Smith. The rocks were no longer treated as mere mineral masses, but as documents from which the detailed history of the earth and its inhabitants was to be compiled. The remains of plants and animals now took the place of importance which mineral species had formerly held, in so

[1] *Proc. Geol. Soc.*, vol. i. p. 386.

much that they gradually monopolized to themselves the
term "fossil," which, in earlier days, had been given indiscri-
minately to every mineral substance taken out of the earth.
Appeals were made on every hand to living nature as a
guide to the changes of past time.   Zoology and botany
became as essential to the geologists of this younger creed
as mineralogy and chemistry had been to their predecessors.
And thus in a few years, from being a mere subordinate
branch of mineralogical inquiry, accused, and not altogether
unjustly, of indulging more in crude speculation than in
sober observation and induction, geology had sprung into a
foremost place among the great divisions of natural science.
This rapid change could receive no fitter acknowledgment
than in the words of Herschel, who said that in the magnitude and sublimity of the objects of which it treats,
geology ranks next to astronomy, and that at length it was
brought effectually within the list of the inductive sciences.

In the midst of this glow of fresh thought and of vigorous
and ever broadening research, Macculloch's *System* made
its appearance like the sullen protest of the last high-priest
of a supplanted religion.   Few had earned a better claim
than this author to the respect of English geologists for hard,
shrewd, original work, carried on among some of the least
accessible tracts of the British islands, and described at times
with a vigour of pen which not many of his brethren of the
hammer could equal.   He might well have been content to
rest his reputation upon that early work.   Owing perhaps
in large measure to bad health, acting upon a temperament
naturally sensitive, he seemed to regard Scotland and the
older rocks generally as a kind of geological preserve of his
own, over which, though he had for many years retired from

JOHN MACCULLOCH, M.D.
From the Engraving of the Portrait by R. B. Faulkner.

field-work, he could not brook that any one should wield a hammer without some licence from himself. Murchison and Sedgwick had laid themselves open to his wrath by their unauthorized raid into his territory. He made no sign at the time; but a few years afterwards, viz., in 1831, he threw this *System* at the heads of his rivals, and in the face of the geological world. The book may be looked upon as almost the last expiring effort of the old mineralogical school of geology in Britain. In perusing it, the reader might suppose himself to be in the midst of the literature of the end of the previous century. Fossil remains are ignored, together with all the new lines of inquiry which they had opened, and the rocks are described according to their mineral characters, precisely as if William Smith had never lived. And yet the author assures the world that he had kept his manuscript beside him for ten years, " in the hope that some better man would stand forward to represent geological science as it is : but he grieves to say that, during that long period, geology has scarcely received a valuable addition, and not a single fundamental one." As President of the Geological Society, it was Murchison's duty to repel this statement, and to point with just pride to the Transactions of the Society as a monument of what had been done during those ten eventful years.[1]

[1] He does not specially refer to Macculloch's treatment of his own work and that of Sedgwick. But no one can read the *System* without encountering passages which evidently refer, in by no means a complimentary tone, to the two fellow-labourers among the Scottish Red Sandstones. Macculloch's ill health and acrimony seemed to increase with his years. In his last work,—a pamphlet to accompany his Geological Map of Scotland (1836),—published unfortunately after his sad and sudden death, his allusions became even more personal. (See, for example, the last sentence on p. 94, where he refers to " the very ignorant and hypothetical

One of the time-honoured customs of the Geological Society was then, and still is, to hold a dinner on the evening of the anniversary; so that, after the President has given an exhaustive, and sometimes rather exhausting, address in the afternoon, he takes the chair and makes after-dinner speeches in the evening, surrounded with a goodly gathering of geologists and friends, who are of course all agreed as to the great importance of the Society, and the unabating interest of the science which it cultivates. In performing this function Murchison seems to have been so well satisfied with the success of his first public geological dinner that he took some trouble to get it reported in the London papers, and even wrote to a friend in Inverness to secure a notice of it in one of the northern journals !

"The summer of 1832," to quote from his journal, "was begun with the Oxford meeting of the British Association, and of this I need say nothing more than that, under the presidency of Buckland, the body was then licked into shape, and divided into six sections. As the mass of the great guns of the metropolis had now joined us, and also Sedgwick, Whewell, and the best men of Cambridge, our success was assured. Altogether it was (thanks to its proposer, Daubeny) a most auspicious meeting,—the more so as it terminated with an invitation, for the next year, from Cambridge, with my dear colleague Adam Sedgwick as *præses*.

persons.") He speaks of his own labours as completing the geological investigation of Scotland, there being nothing further to be done save what could, after a few weeks of experience, "be effected by a surveyor's drudge, or a Scottish quarryman" (p. 17). So far as Sedgwick and Murchison were concerned, there was no cause for this hostility ; for, though they had differed from him on some points, they had never ignored the great services rendered to geology by Macculloch.

" The remainder of the summer was entirely devoted to researches amidst my new loves, the ' Transition Rocks,' not only by revisiting the old ground to complete my sections, but by greatly extending my survey. I had now determined to set to and map out the region. But, alas! the Ordnance maps of a large portion of the country I had determined to examine were only in the course of construction, or not begun. But I got hold of every scrap I could from the Map Office, then directed by Colby, or from my friend Major Robe at the Tower, and so I set to work in the *terra incognita* to which I afterwards (1835) applied the name of Siluria."

If it be true, as Bacon asserted, that " writing maketh an exact man," it is no less true that mapping makes an exact geologist. Without this kind of training, it is not easy to grasp accurately the details of geological structure, and hence the literature of the science is sadly overloaded with papers and books which, had their authors enjoyed this pre-liminary discipline, would either not have been written, or would at least have been more worthy of perusal. Murchison wisely resolved not to trust merely to eye and memory, but to record what he saw as accurately as he could upon maps. And there can be no doubt that by so doing he gave his work a precision and harmony which it could never have otherwise possessed, and that, even though still falling into some errors, he was enabled to get a firmer hold of the structure of the country which he had resolved to master than he could have obtained in any other way. For, to make his maps complete, he was driven to look into all manner of out-of-the-way nooks and corners, with which, but for that necessity, he might have been little likely to make acquaintance.   It often happens that in such half-hidden places—the course of a mountain torrent, the bottom of a tree-shaded ravine, the gully cut by the frosts and rains of centuries from the face of a lonely hillside— lies the key to the geological structure of the neighbourhood. In pursuit of his quest, therefore, the geologist is driven to double back to and fro over tracts never trodden perhaps by the ordinary tourist, but is many a time amply recompensed by the unexpected insight which this circuitous journeying gives him into the less obtrusive beauties of the landscape.

Though Murchison had already learnt something of the devious nature of a field-geologist's path through a country, he had never before tried anything on so detailed and extensive a scale.   At one time he might have been seen measuring sections in Shropshire; soon thereafter, led on by the rocks, he had got away west into Pembroke. Thence, following up his game, he tracked it through the wilds of Montgomery and Radnor, or south to the hills overlooking the great Welsh coal-field, and back again into the English borders.   For weeks and months together this work went on.   Much of the ground proved difficult to unravel, and cost its explorer many a restless night, for he had now got his head so full of grauwacke, transition rocks, and Old Red Sandstone, that he seems to have been able to think or dream of nothing else.   From his notes, however, we may conjecture that though his days were given to hard work out of doors, the evenings were often pleasantly spent under the hospitable roof of the country gentlemen of the region, some of them old friends, who still enjoyed a quiet joke over the enthusiasm with which he now hunted " grauwacke " instead of foxes.

November, with the opening of the session of the Geological Society, brought him back to London and the usual routine of town life in winter. To Sir Philip Egerton he writes immediately after his return, full of excitement over the summer campaign:—" I have done a fine stroke of work. I have coloured up all the Ordnance Maps I could procure, describing a zone of about twenty or thirty miles in breadth, from the Wrekin and right bank of the Severn to

FOSSILS FROM THE GRAUWACKE (CARADOC ROCKS).

1. Calymene Blumenbachii. 2. Homalonotus bisulcatus. 3. Phacops truncato-caudatus. 4. Tentaculites Anglicus. 5. Lingula crumena (Llandovery). 6. Orthis testudinaria. 7. O. vespertilio. 8. Strophomena tenuistriata. 9. S. grandis. 10. Bellerophon bilobatus. 11. B. nodosus. 12. Orthonota nasuta. 13. Nebulipora lens. 14. Diplograpsus pristis. 15. Graptolithus priodon.

the mouth of the Towey, and I hope to show you four or five distinct natural fossiliferous formations of great thickness in our neglected 'grauwacke,' in which I have got abundance of fossils—many quite new; indeed, I have fished some out of the genuine Old Red Sandstone which overlies all my system. I had a most delightful tour, despite certain premonitory choleritic attacks, which disabled me occasionally. My wife met me in Somersetshire, through which county and Wiltshire and Hants we returned, making visits to old friends till we reached our county near Petersfield, where in the month of October I laid low about sixty brace of cock-pheasants. We reached town on the 6th of this month to open the geological campaign.

" Mantell has discovered great part of a *nov. spec.* of large Saurian in the Weald, which he supposes to be his *dear* Iguanodon, of which you know he never as yet found more than the head and teeth. His paper thereon is to be read next meeting (December 5th), after which I am going down to a battue at Up Park."

From the mass of letters which he allowed to accumulate from month to month, some idea can be gathered of the multifarious and distracting calls which were daily made upon his time and attention during the years of his Presidency. The undisturbed early hours before breakfast are given up to the elaboration of his notes. The morning post brings perhaps, among other epistles, a wail from some country geologist, because he has heard no tidings of an elaborate memoir which he had sent up to the President, in the confident belief that it would at once exercise the collective wisdom of the Society. In the forenoon he has to attend a meeting of committee for securing Abbotsford to Sir Walter's family ; or of another committee which is busy organizing a subscription for a suitable memorial to Cuvier. Then he goes by appointment to meet Chantrey, who had made a design for the Wollaston medal. In the afternoon he may have purposed to get some of his Welsh notes into order; but a foreign geologist, with letters of introduction from some of his friendly Continental brethren of the hammer, appears at his door, whom, after giving up an hour or two to him, he finally takes to Somerset House and consigns to the courtesy of the respected Curator, Lonsdale. In the evening, unless, as often happened, he had engaged himself to dine out, or to hold a geological reception at home, he could attend to his correspondence, or, if that had been already accomplished, he might snatch a few hours to prepare an account of his labours in the field for the Society, his wife at his side preparing his drawings and otherwise aiding in the work.

And yet, despite these numerous avocations, time and opportunity were both found for a flight now and then from the bustle of London to the field-sports and friendly intercourse of a country house. Witness the following account of himself, written on 22d January 1833 :—" I met my wife on my return from Cheltenham, and we paid a visit of a week to Lord Milton, in Northamptonshire, and I must say that I never enjoyed a winter week more. He gave me a mount on a capital thoroughbred, son of Cervantes, but the day was unlucky. It was a woodland fox found in the Bedford purlieus, which took us right into the heart of old ——'s preserves, where the Earl and his Christmas friends were dropping the long-tails. You must excuse me if I say that the ex-Minister in his threadbare tartan, patch over his eyes, hat twisted up behind, on a cock-tailed pony, with large gambadoes, distressed as he was by our irruption, looked a perfect pattern for H. B. to realize the ' ould constitution' of Dan O'Connell. But the distress of the day was the death of a poor whipper-in. I am now writing seven or eight hours per diem, nay, even ten and twelve, to make up for lost time, and to enable me to take the last week of the best shooting in England at Up Park. So you see I am living a very sporting life for a

P.G.S. I am delighted you are coming to the anniversary. Greenough is to be my successor."

The continuous writing to which he refers was required for the preparation of the presidential address at the forthcoming anniversary in February. In looking back over the pages of that forgotten document, we meet with notices of several landmarks in geology, showing in what an eventful period of the history of his favourite science the life of the writer had been cast. Among the names of those whose recent deaths he had to chronicle, and whose deeds it was his duty to record, were Sir James Hall and Cuvier—the one standing at the head of physical geology, and linking that generation with the early glories of the Huttonian school; the other acknowledged to be the great master of that newer school of palæontology which had so greatly altered the aspect and the aims of geological inquiry. Among the topics of then recent discussion, he alludes to the erratic boulders (" foundlings," as the Swiss have called them), which, strewn over the plains of Europe, were beginning to attract attention as evidence of some flood from the North— the first beginning of the deciphering of that wonderful chronicle which has laid before us at last the story of the Ice Age in Europe. Among the announcements of new work he gives a sketch of his own labours among the old rocks of the West, and alludes to those of Sedgwick. But his most important item on this head was the reference to the foundation of the Geological Survey, that great national undertaking, over which, some two-and-twenty years later, he was himself destined to preside, and in charge of which he spent almost the last sixteen years of his life. Very modest was its earliest equipment. Mr. Henry de la Beche had been appointed, in connexion with the Ordnance Survey, " to affix geological colours to the maps of Devonshire, and portions of Somerset, Dorset, and Cornwall." To the tact of that sagacious man the Survey owed its existence, and to his energy and skill it is indebted for its present importance, and the great work which it has so far accomplished.

Writing late in life, and looking back upon this early part of his scientific career in London, Murchison penned the following reminiscences :—" During all these years, viz., 1826-38, I inhabited No. 3 Bryanston Place, and, though I had but a small establishment, I saw very agreeable society, for, independent of my scientific friends, I was visited by men in public life, as well as by the lovers of science, letters, and the arts. With Hallam I was in constant intercourse, and also with Lockhart, and with both of these very different men I kept up an intimacy to their death. When Lockhart came to London every one was afraid of the author of *Peter's Letters to his Kinsfolk*, the more so as the Whigs were rabid against him ; but with intimacy his reserve wore off, and I declare that, amongst my friends, I never knew one who was more lively, amusing, and confiding in dual converse, nor one whose loss I more sincerely mourned. If he was a good hater he was assuredly a warm friend.

" Shortly after Bulwer came to London I asked him to dine, but did not tell him whom he was to meet. He had just issued his *Paul Clifford*, and, meeting for the first time at my table, Lockhart, who had cut it up unmercifully, the young author took huff (for he was then a proud young dandy), and thought I had done so to annoy him. It required all Chantrey's good-humour to keep the party together.

"Sydney Smith, Lord Dudley, Conversation Sharp, Lord Morpeth, the Parkes (now Wensleydales), Lord Lansdowne, even the sensible and aged Duchess Countess of Sutherland, did not disdain our small parties. Lady Davy rarely came, for she was too exclusive.

"Among the foremost of our intimates was the all-accomplished, sensible, modest, and retiring Mrs. Somerville, who with her jolly good husband the Doctor, then the Physician of Chelsea Hospital, was constantly with us. We also often visited them at Chelsea, and met there Mackintosh, and other leading characters,—Mackintosh in particular being a great admirer of the lady philosopher. It was our pleasure to bring this remarkable woman and Wollaston together, and to gather from them crumbs of the profound knowledge which they unostentatiously let fall.[1] When we called on Mrs. S. in the morning, and found her finishing off one of her fine landscapes, or instructing her daughters in music, we necessarily admired her feminine qualities, whilst we knew she was up to every line of La Place's 'Mécanique Céleste.'

"With these notables let me associate my geological friends Charles Stokes and William Broderip. The former, a stockbroker, was one of the most remarkable men I ever knew, albeit he has left little behind him. Never out of England, and constantly occupied in the city, he gave up his evenings, nights, and mornings to other avocations, was versed in all languages and a proficient in most branches of Natural History. My little sketch of him in my anniversary address to the Geological Society gives but an im-

[1] Mrs. Somerville, in her charming memoirs, gives some particulars of her intercourse with Wollaston. See p. 128.

perfect idea of his versatile powers. He was the bosom friend of Chantrey, who also was his constant companion with us or at the sculptor's own house. Then there was dear old Major Clerke, the editor of the *United Service Journal*, my old Marlow *chum*, and last, not least, Theodore Hook, who first met Sydney Smith at my house,[1] and has often, when very far gone, extemporized his songs to us over the piano. But these things were my passing amusement, and I was pondering all the time upon turning everything into a geological use.

"Opposite men of all parties were intermingled with my scientific cronies, Sedgwick, Buckland, Greenough, Fitton, and others. These parties were really intellectual; but now that I live in a big house in Belgrave Square my grand dinners are dull horrors—and it is only when I can manage to have a small one that I enjoy seeing company.

"I meddled little in public matters or politics, though my feeling was Conservative, and I was one of those who was, I confess, alarmed at the great sweep about to be effected by the Reform Bill. So I attended the debates both in Lords and Commons, and was present at the whole of the last day's debate in the latter, and which did not close till five A.M.

"To resume my recollections of my earliest scientific friends in London: I must specially dwell on the great botanist Robert Brown, who was chiefly to be met with at the Sunday breakfasts of Charles Stokes in Gray's Inn, and who

---

[1] It is said (Timbs's *Lives of the Humourists*, vol. ii. p. 276) that Sydney Smith and Theodore Hook met at table only twice : first at the house of Lady Stepney, where "they were both delightful and mutually delighted;" and secondly, soon after, on the occasion mentioned in the text, where they met in a somewhat larger party, but where poor Hook's failing became only too visible.

provoked my impatient temper because he never would pronounce upon the genus—scarcely even upon the class—of a fossil plant. Profound in his acquaintance with living plants, he knew too well the fine limits and subtle distinctions to be observed ; these being generally obliterated, and the fructification being rarely visible, he paused and looked again and again, and came to no conclusion. Lindley, on the other hand, being of a less cautious temperament, often dashed off an opinion, and therefore gratified geologists. Robert Brown, though a quiet sedate man, was full of dry humour, and told many a good story to his intimate friends, among whom I was delighted to be reckoned till the day of his death. I was one of the mourners at his burial at Kensal Green, when this illustrious man had but a few old friends to pay the last honours. How different was it but the day before yesterday, when the popular novelist was interred in the same place ! Doubtless, so good a master of English, so smart a satirist, so warm-hearted a friend, and so attractive a writer as Thackeray, merited all the eulogy which has been poured out on his character by all the press. But if a man of science dies, however eminent he be, a passing commendation is all he obtains, and it is doubtful whether the compilers of such works as the *Annual Register* will ever think it right to allude to the death of the first botanist of our era. Nor can a different verdict be expected from the masses or the fashionable world. Every one knows *Cornhill* and *Punch*, *Pendennis* or *Vanity Fair*, or some one of Thackeray's good novels, and so that author obtained a good share of the public applause which the nation accorded to Walter Scott, whilst the *Princeps Botanicorum* of Europe dies unknown by English scribes.

"Among my intimates and correspondents of the first years of my geological career I must not omit to mention George William Featherstonhaugh. He has played a bustling and useful part through life, has published on a vast variety of subjects, and was a most lively, agreeable companion. He was the first to introduce our modern ideas of geology into the United States, which he did with great energy in the year 1831. Afterwards he induced General Jackson, then the President, to appoint him 'State Geologist,' in which capacity he made two extensive tours, illustrating them with long sections. . . . In the French revolution of 1848, when Louis Philippe fled from Paris and was hid in a cottage with Queen Amélie on the south bank of the Seine opposite to Havre, it was Featherstonhaugh, then British consul at Havre, who managed to get the family of 'Mr. Smith' over by night, and popped them into a British steam-packet. Even in this act the consul was the geologist, for he passed off the ex-King as his uncle William Smith, the father of English geology!"

CHAPTER XII.

THE SILURIAN SYSTEM.

DURING the tenure of his Presidency of the Geological
Society Murchison had greatly raised his scientific position
in the country, both in regard to power of original geological
work, and to that practical turn of mind and suavity of
manner which fit a man to play a prominent and useful
part among his fellow-men. He hardly as yet realized the
real importance of the field-work which he had been carrying on among the Transition rocks. Very slowly as the
years passed away did he come to see how full of significance were the sections which he had brought to light
along the Welsh borders.

A few weeks after resigning the Chair of the Society he
gave the first detailed account of what he had been doing
during the two previous years among the "transition rocks"
and "grauwacke" on the border-land of England and Wales.
The brief abstract of the paper to the Geological Society in
which these details are communicated contains the first

imperfect and partly erroneous sketch of a classification which has since become so familiar to geologists.[1]

Released from work in town, Murchison sped back to his rocks on the Welsh frontier, and passed the summer of 1833 in constant travel and work among them, "rummaging the country," as he said, in search of fossils and evidences of the order of sequence among the formations. Again his wife became a partner in the tramp, and while he made more distant forays, employed her pencil on some of the sketches which afterwards appeared to such good purpose in the "Silurian System." On one occasion the monotony of "the perpetual cracking of stones" was pleasantly interrupted by the appearance at the inn of that "famous talker, Richard Sharp," who, in taking leave of the enthusiastic geologist, remarked to him, "Well, my good fellow, I feel assured that you will end in becoming Lord Grauwacke."

While increasing his knowledge of the rocks, Murchison managed also to augment his acquaintance with the inhabitants of the country. Not always, however, to the advantage of his scientific pursuits, for, as he used to say later in life, "Good living in an aristocratical mansion is hostile to geological research. I must honestly declare that

[1] The subdivisions may be quoted here :—

" I. *Upper Ludlow Rock*—Equivalent, Grauwacke Sandstone of Tortworth, etc.

II. *Wenlock Limestone*—Equivalents, Dudley Limestone, Transition Limestone, etc.

III. *Lower Ludlow Rock*—Equivalent, 'Die earth.'

IV. *Shelly Sandstones*—Equivalent, ———?

V. *Black Trilobite Flagstone, etc.*—Equivalent, ———?

VI. Red Conglomerate, Sandstone, and Slaty Schist."

Proc. Geol. Soc., vol. i. p. 475.

In this table the Aymestry and Wenlock Limestones are confounded, and hence the Lower Ludlow Rock is placed under instead of above the Wenlock Limestone.

in general I have done twice as much work when quartered
in an inn." It was in such a mansion, however, that a
project took its rise during this autumn, which came in the
end to make one of the landmarks of his life, and at the
same time an epoch in the literature of geology. His friend
Mr. Frankland Lewis had suggested that he should not
be content with the limited circle of readers which perused

View of the Breidden Hills near Welsh Pool, from Powis Castle.
(Sketched by Mrs. Murchison.)

the ponderous Transactions of the Geological Society, but
should appeal to a wider public, and elaborate into a separate
volume his researches among the old rocks of the English
and Welsh border-land. This idea found a warm supporter
in Lord Clive, at Powis Castle, where Murchison agreed
to undertake the task. Before the middle of November
Lord Clive announced to him a list of eighty subscribers to
the proposed work.

"I have truly done much work this summer," he writes to Mr. Phillips, "having been seventeen weeks hammering, with only one day of intermission. But you gallop when you suppose I am ready for the press. Absorbed in your own great undertaking,[1] you have not had time to think of the magnitude of mine. *Imprimis*, My inquiries range over seven counties, and they dive into the arcana of formations of which no precursor has *written one line!* Hence each succeeding year in which I propagate the principles of our craft, and enlist raw recruits in provinces where the sound of the word geology was never heard before, I find on revisiting my fields of battle that my aides-de-camp have collected facts, and facts alter preconceived notions."

And so the work went on from the Vale of Severn to St. David's. The proposed big book could not possibly make its appearance until after far more complete and detailed examination. Meanwhile each summer's labours were duly communicated in abstract to the Geological Society. From his friends there, such as Greenough, Lonsdale, and Phillips, came letters of encouragement which brought the enthusiastic geologist back to London with renewed energy for work. The campaign of the autumn of 1833 ended by the despatch of five boxes full of specimens from the old "grauwacke" of the west to the apartments of the Geological Society. Lonsdale, ever catering for the wants of the Society, looked forward with his quiet glee to ever so many evenings of amusement and instruction to be had out of these boxes and the notes by which they were to be illustrated. We can picture him in his little den at Somerset House surrounded with books, papers, and specimens, rubbing his

[1] The *Geology of Yorkshire*, now a classic work in British geology.

hands as he wrote to Murchison—"Poor old Grauwacke will be cut up piecemeal." Poor old Grauwacke indeed! With the Woodwardian Professor hewing at him in Cumberland and North Wales, and the President of the Geological Society hacking at him all along the Welsh border, his doom was evidently sealed.

"Perhaps no one better than Lonsdale comprehended the true meaning of the work which Murchison undertook. Certainly no one gave more effectual assistance in the often delicate task of clearing up in the calmness of the closet the difficulties which frequently misled the eager enthusiast in the field. Murchison was never slow in acknowledging his great obligation to his patient and right-judging friend."[1]

Mr. Lonsdale's anticipations were fully realized during that session of 1833-4. From the note-books of the previous summer Murchison furnished four separate papers on different parts of the geology of the districts among which he had been at work. One of these contained the first published table of the Transition rocks of England and Wales, in which they were parcelled out into distinct formations, each characterized by a peculiar assemblage of organic remains. The arrangement showed a considerable elaboration and improvement upon that of the previous year.[2]

---

[1] From MS. reminiscences kindly contributed by Professor Phillips.

[2] The subdivisions now adopted were as follows:—

Old Red Sandstone.

Upper Grauwacke Series		
I. Ludlow rocks,	{	Upper Ludlow rock. / Aymestry and Sedgeley limestone. / Lower Ludlow rock.
II. Wenlock and Dudley rocks,	{	Wenlock and Dudley limestone. / Wenlock and Dudley shale.
III. Horderly and Mayhill rocks,	{	Flags. / Sandstone grits and limestones.
IV. Builth and Llandeilo flags.		
V. Longmynd and Gwastaden rocks.		

A characteristic account of those papers and their reception was given by their author in a letter to Sir Philip Egerton (3d February 1834) :—"Though I say it who should not, I must fairly tell you that the season [at the Geological Society] has not yet produced much, except the communications I have made. I judge as much from our friend Lonsdale's estimate as from my own, perhaps perverted, vision. . . . By accident I had a very good dress circle on my second night, for besides Buckland, Warburton, Lyell, De la Beche,

The Caradoc Range. (Sketched by Mrs. Stackhouse Acton.)

and performers who *could* understand it, the President of H.M. Council, the M. of Lansdowne, dined with me at the club, having quitted a Colonial Council to do so, and he sat it all through the evening."

Important as were these communications to the Society, they could only be abstracts of the work of the long summer campaigns. The full details were now to be elaborated for the *opus magnum* on which the energies of the next four years were to be concentrated. By the month of August all the preliminaries as to publication had been arranged with Mr.

Murray, and the forthcoming work was advertised as in preparation. But much still required to be done in the field in tracing out the geological changes in the long strip of country through which the Transition rocks extended. Hence as soon as he could get away from town Murchison buckled on his hammer again, and betook himself to a re-examination of his old ground in Shropshire and adjoining counties. Up till this time Sedgwick and he had been labouring independently among the old grauwacke rocks, as if each had got hold of a very distinct problem which could be, and indeed needed to be, separately solved. The domains which they had seized were conterminous, and tacitly a sort of 'bateable land had been allowed to stretch between them. It was in the summer of this year (1834) that they met to arrange, if possible, an amicable adjustment of boundaries. Sedgwick crossed over into his friend's territories to make with him a conjoint tour, which was thus described at the time in two letters from Murchison to Dr. Whewell, dated 18th July:—

"' The first of men' took leave of me and my little carriage at Ludlow, on the 10th July, bending his steps (nearly as firm as I ever knew them) toward Denbighshire. We not only put up our horses together, but have actually made our formations embrace each other in a manner so true, and therefore so affectionate, that the evidence thereof would even melt the heart, if it did not convince the severe judgment, of some Cantab. mathematicos of my acquaintance."

"Having dovetailed our respective upper and lower rocks in a manner most satisfactory to both of us, I hastened back to join my wife. . . . I shall run down to Edinburgh just in time for the meeting, and the feast being over, the

Professor and self intend to look at some other border cases
of transition,—the whole to conclude with a lecture from
him to myself on his strong ground of Cumberland. I was
not a little proud of having such a pupil; and although I
think and hope he endeavoured to pick every hole he could
in my arrangement, he has confirmed all my views, some of
which, from the difficulties which environed me, I was very
nervous about until I had such a *backer*. But I will say no
more of Number One than to assure you that we had a most
delightful and profitable tour in every way, and that our
section across the Berwyns, in which the Professor became
my instructor, was of infinite use to me. Such are the foldings and repetitions that my 'black flags' of Llandeilo are
reproduced even on the eastern side of these mountains, and
it is only as you get *into* them that you take final leave of
my upper groups, and get fairly sunk in the old slaty systems
of the Professor.

" I will leave him to tell you of all our marches and
countermarches in Hereford, Brecon, Caermarthen, Montgomery, and Salop. . . . Whether he fell in love with some
of the Salopian lasses or not is in his own breast; but I can
assure you that a whole houseful of them are deeply smitten
with him. When we parted at Ludlow it was found that he
had left that beautiful brown coat of his in the very house
where all these sirens were, so I left him posting back to
recover the old garment, and perhaps to leave his heart." [1]

[1] From this letter it will be seen that Murchison at least was fully convinced of the dovetailing of his groups of rock with the older slaty masses
on which Sedgwick had been at work more to the north and west. As
we shall find, he published this conviction without note or protest from
his friend, who indeed publicly accepted and declared the same belief (see
postea, p. 230). Many years afterwards, however, when bitterness had
arisen between these two comrades, and when perhaps the recollection of

The British Association held its meeting this year in Edinburgh. Thither the two fellow-labourers made their way, the one to resign the Presidency which he had held so successfully at Cambridge, the other to show his Grauwacke and Old Red Sandstone maps, and to take a share in the task of still further consolidating and strengthening the infant Association.

In a letter written to Sir Philip Egerton on his way south again to the Welsh and Shropshire rocks, Murchison thus refers to the doings at Edinburgh, and afterwards :—" The meeting was most successful in every way. . . . I may say,

what actually took place at the time with which we are dealing had become in some measure indistinct, Sedgwick penned and published an account of this first conjoint tour in Wales, differing considerably from that given in the letter quoted in the text.   He says,—" There were early difficulties, both physical and palæontological, in distinguishing the Lower Silurian from the Upper Cambrian groups, and in fixing their true geographical limits, and it was partly in the hopes of settling such points of doubt that in 1834 I went, during six weeks, under my friend's personal guidance, to examine the order of succession as established by himself in the typical Silurian country.   Beginning therefore at Llandeilo, and ending the first part of our joint work at Welsh Pool, we examined many of his best sections.   Occasionally, while he was working out minute details, I spent some days in collecting fossils. . . . I believed his sections, so far as I saw them, to be true to nature ; and I never suspected (nor had he then suspected) any discordancy or break of continuity amongst his typical rocks from the Upper Ludlow down to the Llandeilo groups.   I adopted all his groups, I may say, with implicit faith, never dreaming of a chance (during a rapid visit) of correcting those elaborate sections on which he had bestowed so much successful labour. . . . We never examined or discussed together the Silurian base-line in the country south of Welsh Pool; and whatever be the merit or demerit of the base-line afterwards published in the map of the 'Silurian system,' belongs exclusively to my friend. [See *postea*, p. 307.] As to this base-line, I neither gave nor had I an opportunity of giving any opinion, either good or bad. . . . North of Welsh Pool we reached a country (east of the Berwyns) with which I was previously acquainted. . . . My friend now made use of and interpreted some of my field sections of 1832. . . . I guided my friend (as he in his Silurian country had guided me) over the Berwyn chain to the Bala limestone, along the high road from Rhaiadr to Bala.   We made no mistake in the section. . . . My friend then de-

PROFESSOR JOHN PLAYFAIR.
From a Painting by Sir Henry Raeburn.

without vanity, that we geologicals were all the fashion, and engrossed by far the greater share of attention. Agassiz has pronounced that not one of the fossils of the Burdiehouse limestone are reptiles, but all belong to fishes. You will be amused to read old Buckie's lecture, given two nights before Agassiz made his decision against the reptiles, for in it the reptiles made a grand figure. My fishes in the Old Red are baptized Cephalaspis, from their horse-shoe heads. . . . I was a day at Lord Melville's, after which Sedgwick and self moved on together to Sir John Hall's at Dunglass to look at St. Abb's Head and the Siccar Point, both famous by the writings of Hutton, Playfair, and Hall. Whilst at Dunglass

clared that the Bala limestone was no part of his Silurian system." The Professor points out the error in classifying the Bala rocks as underlying all the Silurian groups, their true place being the equivalent of that of the Caradoc rocks in the lower Silurian series. He asserts that for this error, hardly avoidable at the time it was made, Murchison was alone responsible. It is difficult to see on what evidence this charge rests. One fact at least is certain, that if Murchison started the error, Sedgwick adopted it and believed it for years, although, according to his own showing, the means existed in his own territory of putting the matter to rights at once. "A single traverse from Glyn Ceiriog to the northern end of the Berwyn chain would have settled this question on evidence not short of a physical demonstration. But we did not make this traverse."—*British Palæozoic Fossils*, Introduction, pp. xliii-xlv (1855). But evidence may be found in Sedgwick's own letters to show that he thought and wrote under at least the impression that his own Welsh rocks were older than those of Murchison. Thus even so far back as February 1833 he wrote to his friend in reference to a proposed dovetailing of their work :—"The upper system of deposits, with its subdivisions, is as plain as daylight, and entirely under your set." It would be easy to multiply quotations from contemporary geological literature to show that this was the general impression among geologists as to the views of the two pioneers in Wales. As an illustrative example, reference may be made to the first edition of Lyell's *Elements of Geology*, published in 1838, before the appearance of Murchison's *Silurian System*. See p. 464, where Sedgwick is given as the authority for calling Cambrian a vast thickness of stratified rocks, "below the Silurian strata in the region of the Cumberland lakes, in N. Wales, Cornwall, and other parts of Britain." This subject will come up again in later chapters of this biography.

I fell in with my old friend Lord Elcho, who has set up a
very crack pack of fox-hounds, and he so tempted me with
the offer of a mount on his best nag, that I could not refuse;
and I am still suffering from the stiffness incident to this
frolic, not having been accustomed to screw to my seat for
the last ten years. Sedgwick and myself explored the headland together, and in the boat we had with us our host, Sir
John Hall, and Archibald Alison, a clever young Scotch

View of the Cliffs near St. Abb's Head.   (Sketched by Sir A. Alison.)

advocate, who made sketches of the rocks in my note-
book." [1]

Murchison's journals of this period of his life read very
much like the field notes of an active geologist. Personal
detail is wholly wanting, and the gist of the scientific work
has long been given to the world. From the letters which
he has preserved, we can see what a voluminous correspon-

---

[1] One of these sketches by the future historian and baronet was afterwards introduced into *Siluria* (4th edit., p. 149), and is reproduced here.

dence he must have kept up with friends who lived among his grauwacke rocks, and from whom he derived continual assistance in the shape of notes on the geology, and of fossils. He acknowledged, in his published writings, the value of this co-operation, and gave the names of his principal coadjutors. Even the very children of some of his friends were enlisted in his service, and delighted to get away into the quarries to hunt for fossils for him ; and at a time when these fossils had never been systematically collected and described, it may easily be imagined that this juvenile help proved in many cases eminently serviceable.

It was now plain, after all these campaigns, that though many details might be added afterwards, the grand order of succession of the grauwacke had only been made more clear by every new examination.   It had been subdivided into four well-marked formations, each as defined by mineral characters and fossils as any members of the secondary series.   To continue to apply the terms "grauwacke" or "transition" to these distinct fossiliferous formations, as well as to all the old crumpled unfossiliferous rocks, would evidently lead to endless confusion.   They required a special name.   The story of their nomenclature is thus told by Murchison himself :—" At this time I proposed the term ' Silurian,' and it came about in this way.   My friend, the eminent French geologist, Élie de Beaumont, seeing what a clear classification I had made out by order of superposition and characteristic fossils in each descending formation, earnestly urged me to adopt a name for the whole of the natural groups.   Seeing that the region in which the best types of it occurred was really the country of the Silures of the old British King Caractacus, I adopted that name

[Silurian]. I had seen that all geological names founded on mineral or fossiliferous characters had failed to satisfy, and that fanciful Greek names were still worse. Hence it seemed to me that a well-sounding geographical term, taken from the very region wherein the classification had been elaborated, and where every one might go and see the truthfulness of it, was the best."[1]

The first publication of this new name took place in July 1835 in the pages of the *London and Edinburgh Philosophical Magazine.* In a brief article the author gives his reasons for the proposed term, with some improvements of his previous tabular statement, and a woodcut section to show the way in which the rocks are related to each other in their several subdivisions. As the parent of all subsequent Silurian sections, the diagram possesses a peculiar interest : a facsimile of it is inserted on the opposite page.[2]

Before leaving town for the usual summer work in Siluria he headed a deputation to Government to represent the urgent need of a good map of the northern half of the island—a subject which had occupied the attention of the British Association at Edinburgh. Writing in later years of this incident, he remarks, "Spring Rice, the Chancellor

[1] Murchison's extreme anxiety regarding the names to be chosen for his formations, is well shown in a letter of ten large pages which he addressed to Dr. Whewell on 20th November 1834, " as the great Geological Nomenclator," entreating his assistance in improving his tabular list of the grauwacke rocks.

This section shows in a kind of rough general way the order in which the successive divisions follow each other. It is inaccurate, however, inasmuch as it represents a continuously conformable series from the coalmeasures down to the base of the Llandeilo rocks, and places the latter rocks in a violent unconformability upon those of older date. It was the general belief, as already remarked, that the " Silurian" formations described by Murchison belonged to a younger series of deposits than the rocks of North Wales investigated by Sedgwick.

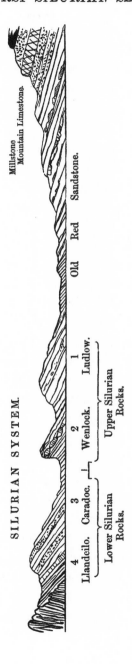

of the Exchequer, received us blandly, and with his Irish
blarney joked me off by saying that I had hunted very well
in Leicestershire with the Melton map, and made several
good shifts.   He avowed, however, that until Ireland got
her map we should or could not get ours.   And so it has
proved."

In its cycle of changes the British Association held its
meeting this year (1835) in Dublin.   We have here no further concern with this assembly than that it was attended
by Murchison and Sedgwick, and that they conjointly gave
there an account of what they had each been doing to
"poor old Grauwacke."   The chief feature of interest about
this communication is the light it casts upon the views
which the two friends entertained of the connexion of their
respective areas of work.   And this becomes a matter of
some importance in relation to the subsequent unhappy
estrangement.   Sedgwick now gave the name of Cambrian
to the rocks among which he had hammered so much in
Cumberland and Wales.   There cannot be the slightest
doubt that at this time, together with Murchison and geologists generally, he regarded his Cambrian masses as older
than the Silurian rocks.   His colleague stated that in
South Wales he had traced many passages from the bottom
of the Silurian system down into the slaty rocks now called
Upper Cambrian.   He himself made no opposition to this
view ; on the contrary, after showing that the lowest Silurian
group was connected with his highest series in the chain of
the Berwyns, he proceeded "to explain the mode of con-
necting Mr. Murchison's researches with his own."   It
turned out in the end that this notion was erroneous, and
that the upper half of Sedgwick's Cambrian rocks was simply

a prolongation of the lower half of Murchison's Silurian. But it was an error in which at this time both of the geologists must be regarded as participating.

The following memoranda from journal and letters give us some notion of the doings of the autumn of this year :—
" —— A frolic in the north of Ireland with Sir P. Egerton, Lord Cole, Sedgwick, my wife, and others, when I made some good geological notes. In clambering along the steep slope [near Giant's Causeway], Sedgwick lost his head, and we much feared that he would fall into the sea. Griffith alone crossed the Devil's Bridge, Sedgwick, Cole, Colonel Montgomery, and self, having turned back and gone up the hill and round. My wife boated all the way and made sketches, and joined us at the comfortable inn of Bush Mills, where we had a very jolly party." Thereafter " I returned to my old hunting grounds of the Silurian region."

" A pleasant visit at Hagly, but I took care to stick to the tail of the Dudley field, which I finished off (ordering a new gun of Westly Richards in a parenthesis). In Tortworth I laboured hard for four or five days, and having completed my map, I then took my departure for Pembrokeshire, sending Madam on to the neighbourhood of Bath to visit some old friends till I became a free man. I spent a day with Conybeare on my road. I then set to work in Pembroke most vigorously, and after three weeks of incessant labour, every day's work proving to me how much I had to do, I left off, perfectly satisfied with having completed a very handsome tail-piece for my Silurians, who are now regularly launched in three bays in Pembrokeshire. What an absurd name does this Grauwacke now appear to be !— I joined my wife two days ago, and shall be in my den to-

morrow, there to shut myself up till the big book is ready—
an awful thought ! "

This self-imposed seclusion would have been serious had
it been carried out, for the big book did not make its appearance for three years afterwards.   The volume grew far
beyond the dimensions originally proposed.   In its preparation, too, questions were continually occurring which made
a re-examination of the ground either desirable or necessary.
Hence, although the winters were spent in tolerably close
application to the desk, the summer months commonly saw
the pen willingly exchanged for the hammer.

As season after season stole past without bringing his
work to light, some even of his geological friends began to
get impatient.   His excuse was thus given to his friend
Phillips in the spring of 1836 :—

" There are at least three reasons why I cannot bring
out the ' Silurian System' with that promptitude with which
you have issued your monograph of the ' Carboniferous
Limestone,'—1st, I have not the same facility of composition.   2dly, I depend on others, and not as you do on yourself, for the description and figuring of the organisms.   3dly,
The work is so multifarious, being, besides the history of the
rocks beneath the Carboniferous system, an attempt to work
out all the general relations of the Lias, New Red Sandstone, and Coal-measures of those central counties. . . .
The work is entirely written save the descriptions of the
organisms—a very large salvo this !   I cannot shove
Sowerby on, and when he is shoved on I am not so sure of
him as I could wish.   My corals I have no doubt will be
beautifully distinguished by Lonsdale; my fishes by Agassiz;
plants I have none ; my graptolites by Dr. Beck of Copen-

hagen. What would I not give, my dear friend, for your powers in the description of the mollusca !"

"The correspondence [with the Council of the British Association on the subject of the delay in completing the Ordnance survey] is ordered to be printed for the use of the House of Commons, who now begin to feel (railroads cutting into their senses) that physical geography is of some importance even to senators."

In such busy but uneventful routine the three years 1836-1839 passed away, the chief feature in each of them being the autumn meeting of the British Association, at which, whether called from the desk or the hill-side, Murchison did not fail to make his appearance.

"In the year 1836," he writes, "I had a good deal of anxiety on account of my dear mother, whose health had been failing, and to whom I had gone at Cheltenham in the spring. This was a cholera year, and my wife having gone down to see her mother at Nursted House, I went in June into Devonshire with Sedgwick to try to understand the complicated geology of that county."

This tour in the south-west of England proved to be the beginning of a series of explorations carried on for three years conjointly by the two geologists, which resulted in the establishment of the Devonian system in geology. For the sake of clearness it will be best to trace out that story by itself in the next chapter. In the meantime we may merely note in passing when and where the explorations were carried on, until we reach the culmination of the Silurian work in the publication of "the big book." It will then be easy and may be useful to turn back for details, and follow out the history of the Devonian

question, which thus to some extent overlapped the
Silurian.

An excellent start was made during this first excursion
into Devonshire, and, as we shall afterwards find, materials
were gathered for a bold announcement to the meeting of
the Association at Bristol. "At that meeting," to resume
our quotations, "the fun of one of the evenings was a lecture
of Buckland's. In that part of his discourse which treated of
Ichnolites, or fossil foot-prints, the Doctor exhibited himself
as a cock or a hen on the edge of a muddy pond, making
impressions by lifting one leg after the other. Many of the
grave people thought our science was altered to buffoonery
by an Oxford Don.

"After the meeting my mother became rapidly worse,
and died at the age of sixty-five, my sister Jeanette and
myself being present in her last sufferings. I buried her in
the same grave with my father, at the little church of Bath-
hampton, near Bath. In the same churchyard my mother's
brother, my old general, Sir Alexander Mackenzie, has since
been interred. No man ever had a more affectionate mother
than myself, her only defect being over-indulgence of her
children."

At the end of the Bristol meeting the Woodwardian
Professor went down into Devon and Cornwall to do some
further hard work among the rocks there. He was at that
time intent on getting a clue to the history of the joints by
which rocks are so abundantly traversed. But Murchison,
having manuscripts and proof-sheets on hand, did not
accompany him, though once again the unwearying explorer
of Siluria found cause to go over part of his old sections to
verify them, driving from town to town in a butcher's cart

which he had hired for the purpose. His friend Phillips met him by appointment at Brecon, to examine with him the curious little tract of Corn y Vaen, and the Professor has made the following memoranda of the journey :—" Welsh ponies were in requisition, and we reached the hill, hoping to escape the jealous company of the Welsh farmer, who looked upon the men of the hammer as some kind of miners secretly prowling for gold or coal. Murchison had paid many visits, and had tried to explain to the inquisitive agriculturist why the barren grey rocks prominent above the 'Old Red Sea' had so much interest in his eyes. On this occasion I also had to encounter ' the old man of the mountain,' because my clinometer was in great use in respect of dip, cleavage and joints. ' Axes of elevation,' ' direction of fault,' ' extent of throw,' ' envelope of old red,' and other strange phrases, made our friend very angry, so that, unlike Welshmen in general, he offered us no kind of welcome or refreshment, but appeared to rejoice in our going away as a relief from some positive evil."

Back in London again among his books and papers, Murchison writes on November 21st to Sir Philip Egerton :—

" I am going through my heavy work, and am just sending to press all that I mean to say of the ' New Red System.' . . .

" My bone-bed in the Ludlow rocks is turning up trumps—jaws with teeth complete, carnivorous shark-like little fellows, with loads of coprolites, indicating that my Silurians digested even harder stuff than your Liassic friends, viz., Pentacrinites, etc. ! This is beautiful, at some 8000 or 10,000 feet below the fish beds at which Buckland begins his transition stories about the oldest fishes. But it will do for his third or fourth edition.[1] He has been in town last week, and was one day closeted with Babbage eight or nine hours, to get his *siphuncle* into order. It appears that Sedgwick and others, on reading the Nautilus Theory, at once saw there was a screw loose in the mechanics, and that if the animal got down to depths unknown he never could get up again. I know not how it is

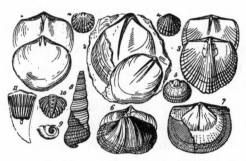

to end, but I hope our friend will be able to sing Resurgam. On the whole the book pleases most people.

" We are going on swimmingly, with bumper meetings. I am working from six A.M. till dark."

Sedgwick had promised to share in the preparation of a memoir on the Devonshire geology, but postponed from week to week the completion of his task. Chafing at this delay Murchison employed a part of the winter in putting together in conjunction with Hugh E. Strickland—then just beginning a career cut short sadly and too soon—a memoir on the New Red Sandstone, in which the English deposits of that age

[1] Reference is here made to Buckland's well-known *Bridgewater Treatise.*

were correlated with the Keuper and Bunter formations of
Germany.   The paper was referred by the Council of the
Society to Sedgwick, and here is his opinion of it as given
to Murchison himself :—

"I have reported favourably on your paper on the
Keuper, and said that it ought to be printed. But was ever
such a blotched, patched, botched, scratched, blurred, bothered
thing sent to an arbitrator ! with a prospectus, too, of certain
plates affixed like a tin case to its tail, I suppose to make it
go.   It made me mutter bad words through my teeth many
times over before I got to the end of it.   Perhaps I did not
swear outright ; but you have no right to tempt me."

This description of the author's style of caligraphy is not
more graphic than true.   His manuscript as it went to the
printers was usually so scored, and crossed, and rewritten, as
to be sometimes with difficulty legible even by himself.
When the proof came back it soon grew under his pen
nearly as bad as the original manuscript, and many a time
had to be set up afresh.   His publisher said of him that he
" wrote in type."

It was in the elaboration of chapter after chapter of
such exasperating manuscript that a good part of the summer
of 1837 passed away.   The affairs of the British Association
entailed indeed a large amount of correspondence and other
duties upon a General Secretary.   The meeting this year
at Liverpool drew Murchison as usual out of his den at
Bryanstone Place, and gave him a week of hard work and
incessant festivity.   For by degrees the rigidly scientific
aspect of the Association had come to be more veiled by
the abundant hospitality and good cheer with which the
members were welcomed.   Each town to which they came

strove to vie with the previously visited places in this non-scientific part of the proceedings. Philosophers, it was found, did not despise a good dinner, and were quite ready to take part in an evening party, or a more formal and crowded soirée.

Liverpool received them on this occasion with the most lavish expenditure. As General Secretary, Murchison had more than enough to do, but he found time to send the following notes to his wife :—

"The preparations here are excellent. *Turtle daily* at the ordinary, so what is to become of the poor savans when they go back to country quarters ? We dine with the Mayor to-morrow, whose lady has a grand soirée in the evening, and thus begin our frolics."

"You are a reasonable woman, and know what a week I have had! nothing have I done but dream, work, and think for the Association. All has gone off admirably, in spite of wind and weather. The conversazione and lighter parties for the evening have been much preferred to the dull affairs of former meetings, and the splendid fête given in the Botanical Gardens to 2600 persons, all of whom were fed, and for which fortunately the day of Friday was fine, contributed no little to the complete success of the thing. Last night we had our finale, and all our thanks."

The rest of 1837, and nearly the whole of the next year were given up to the completion of the "*Magnum Opus*," and the seeing of it through the press, with the drawing and engraving of the map and numerous illustrations with which it was enriched. Not, however, without an occasional malediction over the toil and trouble of the whole enterprise. "I get on slowly and sulkily as respects my own

powers of digestion" (he writes, for example, to Phillips).
" Never will I undertake another big book of such multifarious parts ! But I must now swim through the whole,
or sink under the weight of my own details. I would give
any competent man £100 to launch my ship, but I cannot
trust to others."

The long delay had not been without its advantages in
the greater scope and accuracy which it permitted, especially
as regarded the second half of the work, or that which
treated of organic remains. It had enabled the author
during a series of years to gather the fruits of all the
criticism, the hints, and the information which the discussions of his communications to the Geological Society
evoked. It allowed a steady growth of his geological experience before he should commit himself to the responsi-
bilities of an independent publication, appealing to a wide
circle of readers. Nor had it in any way retarded his
reputation ; for, as we have seen, the more salient features
of his continuous labours in this field, since that lucky
journey in 1831 to the banks of the Wye, had been given
year after year to the Geological Society, and through
the publications of the Society, as well as of the British
Association, had become generally known to geologists
all over the world. But the full account of these, and
notably of the wonderful series of fossils which he had
brought out of the old Transition rocks, had been impatiently
expected for several years. At last, towards the end of
1838, it made its appearance,—a ponderous quarto volume
of 800 pages, with an atlas of plates of fossils and sections,
and a large coloured geological map.

The publication of *The Silurian System,* for so the work

was entitled, forms one of the land-marks in the history
of geology.   It gave, for the first time, a detailed view of
the succession of the geological formations of more ancient
date than the Old Red Sandstone, with full lists, descriptions, and figures of the animals which had peopled the
waters in which these early deposits were laid down.   It
opened up a new chapter, or rather a whole series of chapters, in that marvellous history of life which geology unfolds.

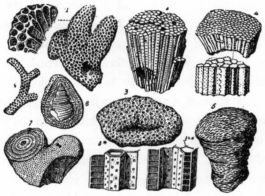

CORALS, ETC., FROM THE GRAUWACKE LIMESTONE (WENLOCK).

1. Favosites cristatus. 2. F. Gotlandicus. 3. A variety of this coral. 3*, 3**. Magnified
portions of two varieties. 4. Favosites asper. 5. Alveolites Labechii. 6. Ceriopora oculata.
7. Favosites fibrosus ; 8. a variety encrusting shells.

Before the researches began, which found their fitting termi-
nation in this splendid work, men had very generally looked
upon the "Transition" rocks as a region of almost hopeless
confusion.   Murchison had succeeded in making out the
order of their upper and most fossiliferous portions, and now in
his pages and plates the subdivisions of these ancient formations stood as definitely grouped and arranged as the orderly
undisturbed Secondary deposits of central England.   He
had traced out also the sites of some of the submarine volcanoes of those early ages, and the great thickness to which

the volcanic detritus had accumulated over the sea-bottom.[1]
To give completeness to his account of the Silurian region,
he had likewise undertaken detailed examinations of the
overlying rocks, including the coal-fields and the various
formations up into the Oolitic series. The results of all this
work were now included in his volume. Rich, therefore,
in original research, and amply illustrated, the book well
deserved the encomium of the President of the Geological
Society (Dr. Whewell), who spoke of it, in his address, " as
an admirable example of the sober and useful splendour
which may grace a geological monograph." [2] No more
remarkable proof of the value of steady industry had for
many a year been given than was furnished by the gradual
elaboration of this work. " If the young student of geology,"
so said a writer at the time, "wishes to find an example of
the effect of diligence and perseverance, as insuring ultimate
success, he cannot do better than to follow the history of the
'Silurian System.'" [3] It was appropriately dedicated to
Sedgwick.[4]

[1] This had been already done in Cumberland by Sedgwick among
rocks then supposed to be older than any part of Murchison's groups,
but which are now known to lie on the same Lower Silurian horizons as
those of Wales. See *Proc. Geol. Soc.*, i. p. 400.

[2] *Proceedings of Geological Society*, vol. iii. p. 81.

[3] *Edin. Rev.* clxvii. 16.

[4] But for the assistance of friends and fellow-labourers, the *Silurian
System* would have been a very different work from what it is. Sedgwick revised some portions of it, especially the Introduction, which he
induced the author in great part to re-write. Agassiz, Sowerby, and
particularly Lonsdale, named and described the greater part of the fossils,
while other friends, whose names are cited in the book, lent a helping
hand. But besides these coadjutors in the preparation of the volume,
the author had been zealously assisted, as we have seen, by active and
disinterested friends in the field, who had worked for him year after
year, and who carried on a voluminous correspondence with him. The
names of some of his coadjutors have been already given. He has himself

During all these busy years, when the author of the
Silurian System was elaborating his work, and giving from
time to time narratives of his progress in the publications
of the Geological Society, the fame of his labours had spread
into every quarter of the globe where geology was cultivated.   His term "Silurian" had been adopted and applied
to the rocks of different countries where similar groups of
fossils were found.   Thus Élie de Beaumont and Dufrénoy
in France, Boué and De Verneuil in Turkey, Forchhammer
in Scandinavia, Featherstonehaugh and Rogers in America,

referred to them in the pages of his work.   But the confession of his
general obligations conveys inadequate ideas of the untiring zeal and
quite incalculable service of some of these friends.   The Rev. T. T. Lewis,
of Aymestry, deserves especially to be had in remembrance, for, without
his generous and effective aid, both in the field and in long and admirable
expository letters, so full a harvest of results could not have been reaped
by Murchison, but must have been shared by other and later labourers.
(See *Edin. Review*, loc. cit.)

In the MS. memoranda already referred to as kindly supplied by Professor Phillips, he says, "Murchison found in Mr. Lewis a man equal to
himself in field-work, and already master of all the local geology.   I had
seen Mr. Lewis's collection in 1836, and often heard his praise from the
Silurian Chief; but by some forgetfulness the record in the great work,
to the foundations of which the Vicar of Aymestry had contributed perhaps more than any other man, was less full and emphatic than might
have been expected."

On the publication of the *Silurian System*, its author showed an anxiety
to have the work favourably reviewed, hardly worthy of his position.
He wrote, for example, an urgent appeal to Sedgwick to pen a criticism
for the columns of the *Times*, and afterwards another entreaty for an
exhaustive article in one of the quarterlies on the whole subject of the
older fossiliferous rocks, the grounds of the request being variously based
on the need of trying to regain some of the large amount of money which
had been expended upon the publication, on the desirability of showing
how necessary a knowledge of geological structure is for the development
of our mineral resources, on the good to geology which might be done by
making the ordinary reading public familiar with some of the more recent
researches, etc.   Sedgwick in a very candid and friendly way assured
him that the book needed no artificial aid, and should be allowed to make
way on its own merits.   Fitton wrote the review in the *Edinburgh*, and
drew attention to the important co-operation of Mr. Lewis.

had accepted his classification, and recognised Silurian fossils in widely distant regions. Hence the book, welcome and long-expected as it undoubtedly was, lost perhaps a little of the novelty which it might otherwise have possessed.

We have now traced Murchison's career up to the completion of the great work of his life. His subsequent geological labours chiefly sprang out of these seven years' toil among the "Transition" rocks. He went abroad to extend the area of his Silurian formations, and he succeeded in achieving its further increase at home. His domain of "Siluria" became, in his eyes, a kind of personal property, over which he watched with solicitude. Or, it might rather perhaps be compared to a vast business which he had established, of every original detail of which he was complete master, and which he laboured to extend into other countries, while he kept up through life a close correspondence with those by whom the foreign extensions were so abundantly and successfully carried out. How all this was done remains to be told in the succeeding chapters.

CHAPTER XIII.

THE DEVONIAN SYSTEM.

WE have now to trace how it came about that another chapter was added to early geological history. With the view of following intelligibly how far this addition was due to Murchison's labours, we may profitably take here a brief retrospect of the previous progress of discovery and opinion regarding the rocks from which the new chapter was compiled.

It was one of the merits of the Wernerian geognosts to point out some of the more salient subdivisions in which, by means mainly of mineral characters, the rocks of the earth's crust may be chronologically grouped. They recognised that their "Transition" series was often covered by red sandstones and conglomerates, and that a younger group of similar sandstones was found to rest upon magnesian limestone or coal.[1] It was in England that this distinction came to be most clearly perceived, because the extensive coalfields of this country were found to separate the two series of sandstones. Hence the terms Old Red Sandstone and New Red Sandstone acquired an important economic signi-

[1] It would appear, however, that the Old Red Sandstone of Werner himself agrees with a part at least of what is now called Permian.

ficance apart from their geological meaning, inasmuch as the one lay below the coal, while the other lay above it.

The Old Red Sandstone during the first quarter of this century had been recognised over a large part of Britain. It was known to occur in broken bands from the Bristol Channel up northwards through the border counties of England and Wales. It had been recognised coming out from under the Carboniferous Limestone in the Lake country. It had been followed for great distances through the Lowlands of Scotland, and along the flanks of the Highlands.

But though the existence of these red sandstones and conglomerates had been extensively proved, little had been gathered regarding their thickness, their subdivisions, their fossil contents, and the general geological history of which they are the records. In Scotland much good observation had been made by Jameson, Boué, Macculloch, Imrie, and others. In England a threefold subdivision of the series was proposed by Buckland and Conybeare.[1] But these rocks were still regarded as only a subordinate, and by no means important, group, being by some geologists placed in the Transition series, and by others with the Carboniferous deposits.

A great advance was made by the conjoined labours of Sedgwick and Murchison among the Old Red Sandstones and Conglomerates of the north of Scotland. They showed the great thickness and importance of the series, its range even up to the most northern parts of our islands, and the great abun-

[1] *Trans. Geol. Soc.*, vol. i. (2d series), p. 210. See also Weaver, *op. cit.* p. 338.

dance and remarkable character of its fossil fishes.[1] It was
therefore with much previous acquaintance with this geological group, that Murchison, in 1831, had begun to trace out
its development in South Wales and the adjacent parts of
England. The vast depth and the variety of strata which it
exhibited in that region, taken in connexion with its extent
in Scotland, had so impressed him with the importance of
the Old Red Sandstone, that when he published the
Silurian System, he proposed, for the first time, to raise it
to the dignity of a distinct geological System.[2] He pointed
out its well-marked lithological characters and its peculiar
fossil treasures as grounds for clear separation. By his
successful search, aided by that of Dr. Lloyd of Ludlow, and
other observers, the fact was made known that the Old
Red Sandstone of England, previously supposed to be
singularly barren of organic remains, did really contain a
number of peculiar fishes, and among them some of the
very same species which had been found in the Old Red
Sandstone of Scotland. By this evidence he was entitled
more confidently than ever to group these rocks of the
United Kingdom in one great series, and when he found
that in South Wales they attained a thickness of nine or
ten thousand feet, he very justly insisted on their claim to
an independent place in the geological record.

These views, however, met with little acceptance on the
Continent. It was objected that with some trifling ex-
ceptions, as for instance in Belgium and perhaps in Russia,
the so-called Old Red Sandstone of the English geologists
did not exist on the mainland of Europe, and therefore that
it had no claim whatever to rank as a system, but could be

[1] See *ante*, p. 144.   [2] *Silurian System*, p. 169.

regarded at the best as a remarkable but only local and abnormal development of the upper Transition or lower Carboniferous strata. There certainly seemed a good deal of force in these objections, and still more in the assertions which were confidently made, that the lowest rocks of the Carboniferous series were found on the Continent passing down into the Grauwacke, and that there was likewise a blending of their respective fossils. If these assertions were well founded, they proved the absence of any intermediate system on the Continent, and rendered the claims of any local British series to rank as a system more than doubtful.

Such, in brief, was the state of this branch of geology at the time of the publication of the *Silurian System.* While the researches out of which that work sprang were still in progress, and the book itself advancing through the press, its author, as already mentioned, was led to begin another series of observations, which led eventually to an important change in English, and indeed of European geology, and to the willing recognition of that "Old Red System" for which contention had in vain been held before.

It was in the year 1836 that the observations now to be followed began to be made. They were the conjoint task of the two long-tried friends Sedgwick and Murchison. Up to that time these geologists had been at work contemporaneously but independently among the older rocks, and though Dr. Whewell, from the chair of the Geological Society, spoke of their labours as "on all accounts to be considered as a joint undertaking," still in actual fact the two pioneers had started from wholly different points, and had, as we have seen, toiled to cut out each his own pathway through that vague and unknown region of "Transition" rocks, which certainly seemed wide enough to give them ample room for exploration without much risk of trenching upon each other's ground. Sedgwick had grappled with the physical structure of the rocks, and, amidst enormous difficulties, had achieved success. Murchison, on the other hand, had found a series of strata where the physical structure was comparatively simple, and which yielded such abundant store of fossils as to be capable of subdivision by their means. But now, in the south-west of England, the two friends were to combine their methods, and to work out a difficult region by help both of physical structure and of organic remains.

There was no such ambitious plan before them, however, when they began their work. They had one definite point to settle, viz., the age of the Culm-measures of Devonshire. But in putting that matter beyond dispute, they were gradually led into further and wider explorations, not in Devon and Cornwall merely, but over a considerable area of the Continent. It was by means of these labours that the "Devonian System" of rocks was established. How the work first took shape is best told in Murchison's own words :—

" The origin of this joint survey [of Devonshire] came about in this way. In the preceding winter,[1] Mr. (afterwards Sir Henry) De la Beche had sent up specimens of small fossil plants from the culm rocks of North Devon, which he described as belonging to the Grauwacke formation. At the evening meeting of the Geological Society I opposed this view, on the ground that my Silurian rocks,

[1] December 1834. See *Geol. Soc. Proc.*, vol. ii. p. 106.

both upper and lower, contained no land plants whatever.[1]
Moreover, I thought I recognised a complete similarity between these common specimens of North Devon and those
which I had explored in the opposite coast of Pembroke,
and which I knew were superposed to the Millstone Grit
and Mountain Limestone. I therefore urged Sedgwick to

---

[1] It is perhaps hardly worth while reverting, even in a foot-note, to a
personal matter which threatened to bring about a rupture of friendly
relations between geologists all of whom have made their mark in the
scientific history of their time, and who are now gone to their rest. And
yet the expressions in the text seem to require further explanation, more
especially as some of the survivors of that time may still be under the
belief that De la Beche was hardly used in this affair. It was asserted
by some of his friends that Murchison and Sedgwick had obtained possession of an early unpublished copy of his Ordnance Geological Survey map
of Devonshire ; that they had, unknown to him, gone down into his territory and examined his sections with the map in their hands ; that they
had thereafter hurried up to the Bristol meeting to make an attack upon
him and expose his mistakes ; and that afterwards, although their full
conjoint paper had not been read to the Geological Society, they procured
a statement and recognition of their views in the anniversary address of
the President. The real facts were these :—When De la Beche announced
the discovery of plants of Carboniferous species in the "Greywacke" of
Devonshire, Murchison (as stated above) opposed this alleged discovery,
because it ran directly counter to all the evidence he had obtained in his
own Silurian domains as to the disappearance of Carboniferous forms of
life from the older rocks, and, as he wrote to De la Beche, "I could not
bring out my long-projected work with such a geological contradiction in
my face." De la Beche invited him to examine the ground for himself,
and gave him directions what to see, and where to see it. The map was
purchased in 1835 in the ordinary way from a bookseller's shop, where it
was sold also to other members of the Society. But it was not used on the
ground until the summer of 1836. Possibly, in the meantime, De la
Beche had begun to suspect the accuracy of these early impressions of
the map. When Sedgwick and Murchison came to the ground, they found
the facts to be as stated above. The supposed "Greywacke" turned out to
be merely a somewhat abnormal condition of the Coal-measures, and, instead of occupying an anticlinal area, so as to dip under the other rocks,
actually lay in a great trough above them. So far De la Beche was undoubtedly wrong, and his opponents were undoubtedly right, as was afterwards shown by the alteration of the Survey map in accordance with the
newer views. The charges of unfairness appear to have been whispered
about by De la Beche's friends in London, while he himself was busy in

join me in a campaign to settle the question.[1] He agreed to
do so. So off we went; and first we looked through the
rocks of North Somerset, Ilfracombe, Morte Bay, Baggy
Point, and Barnstaple. As we went on, a good, steady,
southerly dip continued until we reached the edge of the
famous Culm tract, into and under which the older strata
pitched at a rapid inclination. I there saw that the game
was won, and, drawing a section, in which I reversed De la
Beche's hypothetical diagrams, I called out to Sedgwick
from the rock on which I was sitting,—' Here it is ! Look
at my section of the *North Devon coal-field*—the youngest
instead of the oldest rocks of the county—our job is done !'
Still he was a little incredulous until we advanced southwards (for I had sketched this from the north side of the

the field in the south-western counties. They were indignantly denied at
the time by Murchison, in a letter to De la Beche himself (6th January
1837), and in one to Sedgwick (2d February 1837). That De la Beche
was vexed to find some of the work of the Survey to be wrong was
natural enough, and that Murchison may have shown, as appears from
his narrative above, a little elation in pointing out his friend's error,
was also to be expected. Indeed, it would seem that he allowed himself to
write to Sedgwick in such a way about the alleged discovery of a Grauwacke flora in Devonshire as to call down remonstrance from his com-
rade. Even as far back as January 1835, that is only a month after De
la Beche's announcement, we find him acknowledging Sedgwick's com-
plaint thus :—" You were quite right in reproving me if you thought that
I used any acrimony in speaking of De la Beche's discovery, but I had
long before obviated the possibility of such being the case on one side or
the other by a friendly interchange of opinions with De la Beche himself." But a perusal of the correspondence and of the published papers
and abstracts has convinced the writer of these lines that no unfairness
can be justly attributed either to Murchison or Sedgwick in the matter.
It may be added, that though right as to the relative position they
assigned to the Culm-measures, these authors were much deceived in their
identification of the underlying rocks with the Silurian and Cambrian
systems, as will be shown in the sequel.

[1] In a letter of 8th February 1836, Sedgwick proposes to Murchison
and plans the tour in Devon and Cornwall. It may have been previously
suggested by Murchison.

bay), and then when he saw the actual order he entirely
assented, saying what a crow we should have over De la
Beche. The truth I can only surmise to be, that De la
Beche, who was certainly a very able geologist, had never
really looked carefully at the consecutive sections in nature,
but seeing the Culm strata in a state of great contortion in
a low tract, he had presumed that they passed under the
higher country in the north. I also believe that he was so
much occupied in writing that remarkably skilful and ingenious work (the best he ever wrote), *Theoretical Researches
in Geology*, that in doing so, and carrying out his first map
of Devon and Cornwall, he really worked very little in the
field."

"At the Bristol meeting of the British Association, the
chief business of Sedgwick and self was to establish the
point regarding the great change we proposed in the structure of Devonshire; and though Greenough, Buckland, and
the old hands made some resistance, and did not like to see
the ancient 'Shillats' and 'Gossans,' believed to be the most
ancient rocks in Britain, so modernized, it was evident that
truth would prevail."

After the meeting, while Murchison, as we have already
noted, returned to his literary toil in London, his friend and
coadjutor went again into the Devonshire country, and spent
many weeks in hard work there, so that a broad base was
thereby laid for the conjoint paper which it had been
arranged to read before the Geological Society.

But the conclusions arrived at by Sedgwick and Murchison, though they have now been for many years part of
the common fund of geological knowledge,[1] were far from

[1] The main point established by them was that the Culm-measures lay

meeting with general acceptance at first.   Some idea of the
opposition, or at least of Murchison's estimate of it, may be
formed from the following sentences in a letter to Phillips, of
4th January 1836 :—" The paper by old Weaver was read
last night, and the fight is over.   He has sided completely
with S. and self.   Austen, a remarkably clever young geologist, is also with us ; Major Harding from the first with us.
The case therefore stands thus : For the old constitution—
Greenough, De la Beche, and Parson Williams.   On our side
are the two geologists of Great Britain who have given the
longest attention to the old fossiliferous strata, and their
opinions are supported by every man who has gone into the
tract to judge for himself.

" All the support expected from France has gone against
the ancients ; for Buckland (himself as unwilling a witness
as Weaver) comes back from France persuaded that Élie de
Beaumont's "Grauwacke coal-fields" are nothing but ordinary
Carboniferous deposits reposing on Silurian rocks.

" We are effecting a great reform at the Geological, to
save Lonsdale's life, and enable him to do his quantum of
duty.   We split the duties—Lonsdale, assistant secretary
and editor ; a curator to be found.   R. I. M. chairman of a
committee to find said curator."

The " fight" alluded to in this letter, however, was merely
a preliminary skirmish on the reading of a memoir by
another member of the Society, and though valuable as
giving some notion of the relative strength of the parties,
by no means ended the warfare.   Murchison counted much

at the top of the Devon rocks, and belonged to the Carboniferous system.
On what particular horizon in that system they should be placed does
not appear to be satisfactorily settled yet.

on the support of the Woodwardian Professor, who, if he could only be got into such measure of health and spirits as to come up to town for the purpose, would easily and triumphantly rout the enemy. Thus on the 30th January the following appeal left London :—

" MY DEAR SEDGWICK,—I worked all day yesterday to make the sections, and to have them correspond with our long Bristolian *coupe*. I was in great hopes to have your despatches before now ; but I wait patiently like a lamb for the sacrifice ;—and sacrificed I most assuredly shall be without your aid. However, I will drink the best part of a bottle of sherry to screw me up to face Buckland, Greenough, Yates, and the Ordnance forces which are to be brought against us. In anticipation of the memoir, I must take this chance of a *vale* from you before the fight."

Upwards of six years had elapsed since these two fellow-labourers with the hammer had been leagued together with the pen. The brief notice of their discovery made to the British Association was meant to be merely a prelude to the much fuller memoir designed for the learned audience at Somerset House. Former experience, however, showed that the Woodwardian Professor could not be got to move faster than his wonted pace. After many delays and promises, a date was fixed for the reading of this memoir. Murchison duly appeared, but found neither Sedgwick nor the paper. The letters which came up week after week from Cambridge had brought the most touching lamentations over the exacting claims of lectures, examinations, audits, and other University business, and hardly one of them ever failed to carry a bulletin of the progress of the influenza, gout, dyspepsia, nervousness, or other of the bodily ailments under which the writer happened to be groaning at the time, and which he anathematized with whimsical fervour. Murchison's chagrin was expressed next day as follows:—

"3 BRYANSTON PLACE, *2d February* 1837.

" MY DEAR SEDGWICK,—The part of Hamlet being omitted, the play was not performed, and all the scenic arrangements which I had laboured at were thrown away, though the room looked splendid. The morning's arrivals certainly surprised me. Ten o'clock brought me your double letter; eleven o'clock by the same mail the maps, and a little note to Lyell, but in vain I looked through the parcel for the document to be read. I read and re-read your letter, and still I could not understand it. One thing I clearly perceived, and with great regret, that you were seriously out of sorts, and had been suffering; so after waiting till two, I journeyed down to the Society, still thinking that a third package with the paper might be sent to Somerset House,— not so, however. These things going on; the whole room decorated for the fight; Buckland arrived, Fitton present, and a large meeting expected,—what was to be done? Fitton and Lonsdale . . . counselled me to give up the thing, which I resolved to do, to the very great annoyance of the President [Buckland], and of all the others who came to hear. . . . .

" I am mortified that the memoir did not come; of course I blame myself somewhat for having thrown in doubts on some points, because I see that ill as you have been, and without the power on my part of talking the case over, we mutually misapprehended each other. But enough of what is past. The thing now to consider is when to have the paper out. I should certainly not wish to have it done *till you are present,* because we must have a fair stand-up fight and knock the —— and Greenough down.

" We had a good discussion on Buckland's Keuper, on which Greenough and myself agreed about the absurd term *poikilitic,* backed by old Paddy [Fitton], so the spots were damned. We had a supper at Cole's,—Buckland, Horner, Stokes, the Viscount, Sir Phil, and my friend Rosthorn of Wolfsberg, a great friend of the Archduke John's, present.

" Did you really imagine that I was to dramatize the whole thing without a sermon before me ? or have you been written to by Greenough or some of the dark school ? or was the paper unfit to be sent ? or was it omitted by accident and mistake ? The President stated the last as the cause, and I said not a word about it, for with Lonsdale's help in construing your letter, we were unable to understand it. I think that the delay occasioned by my doubts and your influenza and state of the stomach are the true causes ; but if you had sent it in ever so unfinished a state, the heads would have been read, and an abstract made, which would have served all purposes."

Summer had made some progress before the paper was at last actually read to the Society. It was the first of a series of memoirs upon the rocks of Devon and Cornwall, and their equivalents elsewhere.

The settlement of the geological age of the Culm-measures of Devonshire, though by no means an unimportant question in British geology, was of small moment compared with the further researches to which it led. In working out the position of these rocks, the two fellow-labourers found it necessary to get a base-line for their Carboniferous forma-

256 *SIR RODERICK MURCHISON.* [1837.

tions. In other tracts of this country they would have met
with ordinary Old Red Sandstone. But in Devonshire and
Cornwall they encountered a series of rocks which had
undergone so much alteration that their true position was
difficult to define. They were usually classed by the old
and uncouth term Grauwacke. In some respects they
resembled the old slaty masses of Wales, and at first the two
geologists who had come to them fresh from these Welsh
deposits made them out to be actually in the same geological position as the middle and upper parts of Professor
Sedgwick's Cambrian series of North Wales.[1] A good deal
of limestone, with an abundance of fossil remains, distinguished these Devonshire strata. But owing to the way in
which the rocks had been squeezed and broken, their order
of succession was not easily ascertained.

Various observers, especially Mr. Hennah, Mr. (Godwin)
Austen, Mr. Williams, and Major Harding, had made
collections of the fossils, which certainly differed considerably from those of the Silurian rocks, and quite justified
Murchison in deciding not to claim these strata as part of his
Silurian domain. Mr. Lonsdale, toward the end of 1837, after
an examination of various collections of South Devon fossils,
came to the conclusion that the rocks from which they were
obtained must be intermediate between the Silurian and the
Carboniferous series, that is, on the same general parallel as the
Old Red Sandstone of other districts.[2] He was led to this
inference purely on palæontological grounds, because some
of the fossils belonged to Silurian species, while others had

[1] *Proc. Geol. Soc.*, ii. 560 (June 1837).

[2] *Proc. Geol. Soc.*, iii. 281, and *Trans.*, 2d series, vol. v. p. 721. In this
memoir, the author gives references to previous authors on the rocks of
Devon and Cornwall.

a distinctly Carboniferous character.   This idea, however, was not immediately adopted by Sedgwick and Murchison, for they could not get the Welsh and northern type out of their minds.

While the Woodwardian Professor and the author of the " Silurian System " were still groping their way among the puzzling rocks which underlie the Carboniferous deposits of the south-west of England, another labourer, hitherto unknown, had been for many years collecting and pondering

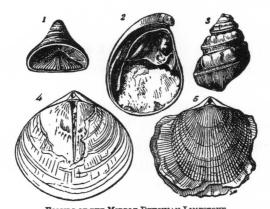

FOSSILS OF THE MIDDLE DEVONIAN LIMESTONE.
1. Calceola sandalina.   2. Megalodon cucullatus.   3. Murchisonia bilineata.
4. Stringocephalus Burtini.   5. Atrypa desquamata.

over the strange fishes which lie entombed in the Old Red Sandstone of the far north of Scotland.   The name of Hugh Miller is now familiar wherever English literature has made its way.   At the time of which we are treating, it had been heard of out of his own Cromarty district as that of a musing, meditative stone-mason, who employed his leisure hours in writing rather indifferent poetry and most graphic and vigorous prose.   In what other pursuits the intervals of his manual labour were spent, and notably how he began to in-

terest himself and others in stones and their story, he has
told in his own charming memoirs. The following letter,
one of the earliest which he addressed to his future friend
Murchison, is characteristic :—

" Cromarty, 1*st June* 1838.

" Honoured Sir,—My friend Dr. Malcolmson of Madras
has written me from Paris, that he has had an interview
with M. Agassiz, and that that gentleman has expressed a
wish to see one of the fossils of a small collection which I have
been forming for the last few years. The Doctor also men-
tions to me in another letter that he had had the pleasure of
meeting with you in London about the middle of last spring,
and that you were at that time engaged in researches which
some of my specimens might perhaps serve to illustrate.
From a further remark, I infer that you too are desirous of
examining some of them. I herewith send a few of the
more portable to Agassiz, requesting him (should he be no
former of collections himself, which Dr. Malcolmson tells
me he is not) to send them to you, who deserve so well of
the geologists of the north, when he has looked over them.
Lest, however, some accident should detain them on the Continent, I deem it proper that you should have an opportunity
of examining them in the passing, and I have therefore
requested Mr. James Malcolmson, the Doctor's brother, to
forward them to your address, with which I myself am
unacquainted. . . . [Here follow some descriptions of
the fossils.]

" There is one question in connexion with these fossils
to which I would fain receive an answer, and which I have
put to Agassiz, but which you, sir, could favour me by
answering much sooner than I can expect to hear from him.

Is the formation in which they occur a fresh-water one, or otherwise ? I have some intention at present of drawing up a popular account of the geology of this part of the country for a widely-circulated periodical to which I occasionally contribute, and the fact in question, if an ascertainable one, is essential to my purpose. Your letter, were you to favour me with a very few lines on the subject, would find me in Cromarty. It would afford me pleasure to forward for your inspection such of my specimens as might prove of use to you in your present researches. I am desirous to make my little collection as complete as possible, and in no place, perhaps, could it be of so much interest as in the middle of the district whose oryctology it illustrates. Some of my specimens, however, are in duplicate, and I need not say how welcome you will be to one out of each of the pairs, and to the use of all the others. Please favour me by sealing my letter to Agassiz ere you make up the box. I do not know that I have addressed that gentleman as I ought, but he must just excuse the ignorance of a foreigner and a provincial in the way the far-famed author of *Salmonia* did the Frenchman who addressed him as *Sirumphrydavy.*—I am, honoured Sir, your most obedient humble servant,

" HUGH MILLER."

From Murchison's reply to this letter a few sentences may be quoted :—

" Although my work was intended to be exclusively devoted to Silurian (or Transition) rocks of England and Wales, I have made a few allusions to other tracts, and, among these, to the Old Red Sandstone of Scotland, in doing which I have, in the descriptions of the organic remains, briefly alluded to your labours.   Now that I know the fidelity and closeness of your research, I shall endeavour to introduce another allusion in the Appendix, which is all that remains unprinted.

" I am delighted with your clear and terse style of description, and beg to assure you, that if you could send us, in the course of the summer, any general and detailed account of both the Sutors, and all their contents, I shall have the utmost pleasure in communicating it to the Geological Society, to be read at the November meeting.

" You write and observe too well to waste your strength in newspaper publications, and a good digest of what you have done ought to be preserved in a permanent work of reference.   I can give you no positive answer as to whether the Old Red Sandstone of Scotland was formed in a lake or in the sea.[1]   I have, however, strong reasons for believing that it is a marine deposit, for in England we find marine shells in it to a considerable height above the uppermost beds of underlying Silurian rocks. . . . I much long to revisit the shores of Caithness and Cromarty with my increased knowledge, and with the conviction that I should learn so much from you, but I fear it is hopeless."

Besides abundant work and correspondence in regard to Devonian geology, Murchison took a leading part in one of the most prominent of the scientific doings of London in this year (1838).   Sir John Herschel, after an absence of four years and a half at the Cape, had returned to England with a rich harvest of astronomical observations.   It was

[1] This question, mainly from the labours of Mr. Godwin-Austen, Professor T. Rupert Jones, and Professor Ramsay, can now be more definitely answered, in a sense opposite to the view which Murchison favoured in this letter.

determined by his scientific and other friends to give him a
public dinner, and to present him on that occasion with an
inscribed vase.   Murchison acted as honorary secretary, and
to judge from the mass of correspondence which remains
among his papers, his post must have left him for many
weeks with hardly an hour of leisure.   One of Herschel's
notes to him concludes " with repeated thanks to you for
all the very great trouble which this affair has caused you."
The gathering proved eminently successful—a result in no
small measure due to the good management of the secretary,
and especially to his facility for grasping even the most
insignificant details, and planning the execution of them.

Before we resume the Devonian story, reference must be
made to the death of Mrs. Hugonin in the beginning of the
year 1838, and to a remarkable letter which that event evoked
from her son-in-law.   This letter is addressed to his " dear
friend" Sedgwick.   It was never sent, however, but remained
in its writer's repositories until his death.   During the interval he appears to have read the letter at least twice—in
1857, and again in 1869—as is shown by his own handwriting on the back.   It would seem, therefore, to have been
regarded as a record worth preserving, of the state of the
writer's mind at the time regarding a momentous subject,
on which, even up to the end of life, he was not given to
speak.   The letter is marked outside in handwriting of a
late date, "My Creed in 1838."

" NURSTED HOUSE, PETERSFIELD,
19th January 1838.

" MY DEAR SEDGWICK,—I have not for the last many
months found an hour so vacant, that if I abstracted it from
the book, or any other avocation, I did not reproach myself,

so heavily has the incubus pressed upon me. Here, however, . . . I am free to occupy an hour, and I give it to you
as the man of my heart. 1st, Talking of this last-mentioned
member of our frame in a physical sense, I must crave some
of that sympathy from you which I have often felt for you
when you have described to me your own sensations in its
region.   The scene here has altogether been trying and
harassing for my wife and self—several times up and down
from town, and, on the last occasion of my visit, I returned
only to Eccleston Street to hurry off Mrs. M. at a moment's
notice, as I feared she would be too late to close her mother's
eyes.   This, however, was happily not the case.   The old
lady made a wonderful rally, her mind became quite composed, and she took the sacrament with her daughter in full
confidence of a change to a better world.   These are agreeable reflections.   To-morrow I attend her body to the grave.
The will gives to my dear wife a most ample income for her
life. . . .

" I do not mean to relax one jot in my search after
natural knowledge ; nay, being now a free agent for the first
time these twenty years, I shall, I hope, be enabled to em-
ploy all my leisure hours more effectively in pursuing my
favourite study.

" But this is not enough.   I have one deep-seated source
of personal unhappiness in my thoughts of the future.   To
go we know not where, may be viewed calmly and resignedly
by many philosophers, trusting as they do to the wise dis-
pensations of Providence, yet *unable* to believe in the great
Atonement for the sins of man.   Alas ! I am (for I need
scarcely confess it again to you, for you know me) one of
those half-instructed wandering beings who sufficiently know

and feel what they *ought* to believe, yet cannot overcome the force of habit and a long-continued apathetic indifference to the vital point. Doubtless I perceive much to admire, nay, nothing to cavil at, in the precepts of Christ, though I cannot bring my mind to acquiesce in His divinity. Still less can I confide in and give my common sense to adopt all the historical details of the Old Testament. You will refer me to Paley, while ———, professing to be a Christian, will refer me to Fellowes. I do not require a stimulus to induce me to adopt natural religion, for I have it strongly implanted in me ; and if geology has done me no other good, it has, at all events, strongly fortified me in this sense.

" But here I halt. Most unwillingly it is true, for few people have a higher respect for *sincere* believers than myself, and no one would more stoutly fight for the Church, as a great and essential moral engine, than myself. When, however, I see men of powerful minds and great integrity, who are strict believers in Christ, I am roused to a perception of the chance there is that the defect is in my own capacity and heart. I hope the former only. Your example has made more impression upon me than all that was ever said or written ; for nothing has more alienated me from Christian belief than the constant exposure (which history and our own experience affirm) of hypocrisy, cant, and all the worst passions veiled under the garb of religion. You might well say to me, ' Look at home ;' for if there ever existed a thoroughly pious, yet unobtrusive Christian, that person is my excellent wife. Seeing the tranquillity with which she views her passage from this world, and knowing how the best Christian principles are ever her guides, albeit without a tincture of fanaticism or exclusive sanctity, I cannot but

hope that the day will come, when, striving to follow out
the dying wishes of my *own* beloved mother, I may become
a true believer.   Alas ! I am a *short* way yet upon the road.
—Ever yours, my dear friend,   ROD. I. MURCHISON.

" Having written, I looked at my confessions and was
about to destroy them, but this would have been giving way
to my own pride : so you must bear with me."

During the winter of 1838-39 Sedgwick and Murchison
were busy trying to get at the meaning of the Devonian
rocks.   Lonsdale's suggestion as to the position of these
strata was now engaging their attention, and they sought
anxiously for light from further fossil evidence.   Many a
box of specimens from Devonshire was turned out and
scrutinized with Sowerby and Lonsdale.   It was not, however, until the spring of 1839 that they quite discarded their
previously published views of the age of the older rocks of
the south-west of England and adopted those of Lonsdale.
Even in March, Sedgwick could still write to his friend,—
" The Devon fossils are a great puzzle ; but I am as firm as
ever—*no Old Red* in Devon."[1]

The two geologists once more became fellow-workers
with the pen.   And the consequence was, of course, a return
to the former kind of correspondence—vehement objurga-
tions by the Professor on his real or imaginary ailments,
with whimsical accounts of his condition, shrewd criticisms
on his friend's writing, and earnest advice as to courtesy and
moderation towards opponents.   The opposition to the re-

[1] Mr. De la Beche's Geological Report on Cornwall and Devon appeared
in 1839, full of excellent observations, but not admitting the Culm rocks
to be true Coal-measures, and retaining his old term Grauwacke for the
older rocks of that region, which were soon to be named Devonian.

form which they wished to effect in the nomenclature of the
older rocks of Devon and Cornwall had not wholly subsided,
for there came now and then a protest or denial from the
other side, though the main point for which they had originally contended—the true overlying position of the Culm-
measures—was now so tacitly admitted as to be claimed as
part of the common stock of knowledge, without reference to
their relation to it as discoverers.   The Ordnance Geological
Survey Report upon the district had just appeared, and irri-
tated them by the way in which it seemed to them to over-
look the important work which they had done in that part
of the country.   They had written and published rather a
sharp retort upon De la Beche,[1] and the atmosphere at the
Geological Society was in that state when a storm such as
had never been experienced at Somerset House might at
any moment have burst forth.   A paper on the Devon
Geology by the Rev. D. Williams, one of the opponents, was
announced for reading on the 10th of April.   A fierce
battle was looked for, and the combatants and would-be on-
lookers came from far and near to be present   Sedgwick
could not attend.   The good fight was therefore left to be
fought by his military ally, who, next day, still full of the
excitement, sent him the following despatch on this subject :—

"11th April 1839.

"MY DEAR SEDGWICK,—The fight is over.   It lasted till
near midnight, and, all things considered, we have come off

---

[1] Reference is made to the paper " On the Classification of the Older
Stratified Rocks of Devon and Cornwall," which had appeared a few days
before in the April number of the *Philosophical Magazine*.   The latter
part of this paper is a rather angry and personal defence of the originality
of their work in these two counties, drawn forth by the statements in
De la Beche's Geological Report on the same district.

remarkably well. Parson Williams, who was present, had prepared an Ordnance map of Devon and Cornwall coloured on his own *mineralogical* plan. . . . Immediately after the memoir was read, De la Beche, who came up per mail for the nonce, rose, and holding in his hand our memoir, commenced an exculpation of himself from the charge we bring against him in our conclusion. . . . . He spoke calmly, and without going into the memoir of the evening. I immediately replied by first assuring the Chair that I had no hesitation in expressing my regret that a word or two had been made use of in the hurry of composition which both of us were sorry for. . . . . Disavowing the least personality, I immediately got D. with me, and having thus cleared the course, I opened the discussion on Williams' paper, and went ' the whole hog,' as well as I could, touching the Devonian case. De la Beche then replied, but did not attempt to shake one of our positions, did not place a veto on one of my assertions, and least of all, on that which laid claim to the originality of the Culm-trough. He bothered about a point or two near Chudleigh, as difficulties, and ending by saying it was immaterial to him what the things were called.

" Lyell then spoke, and very adroitly put the case as one most agreeable to him, now that he perceived that Mr. D. not only acknowledged that the view which we took at Bristol was original, but also that he (D.) was by no means indisposed to adopt our new views, which get rid of all the anomalies and difficulties (about plants and fossils).

" Fitton rose in great solemnity, and with deep pathos impressed on the meeting the propriety of restraining the too pungent expression of controversial writing among geo-

logical friends, alluded to my having called him 'my geological father,' and only wished that I had submitted the
paper in question to his parental revision before it was published. He acknowledged, however, that the explanation
had quite rectified the case, and then he went on to expa-
tiate on the value of our doings, giving us superlative praise,
and bringing out Lonsdale in the foreground.

" Greenough made his oration as I expected, was very
ingeniously sophistical, tried to throw all into chaos, saw
nothing new in our views, adhered to his old belief—Greywacke for ever!—and sustained old Williams by casting
fossil evidence overboard.

" Featherstonehaugh spoke well on the great subdivisions
of the old rocks of North America, and said they were distinctly the same as ours.

" . . . . These and many other things being said and
done, Buckland summed up at half-past eleven, and though
he evidently wished to shield De la Beche, he ended by
approving highly of 'Devonian'—he now saw light—that
light he referred to W. Lonsdale, and henceforth, said he,
there will be two great names in English geology—W.
Smith and W. Lonsdale; he adhered entirely to the fossil
evidences, did not give us the credit we deserved for our
coal-trough (which is the *key* to the whole thing), nor did he
do justice to my Siluriana, without which, as you have justly
said, no one could have started this new hare.

" The room was a bumper. Warburton, who sat it out,
assured me . . . that he looked upon the case as settled, as
it was quite evident that Buckland had completely given in,
De la Beche was ready to do so, and Greenough alone held
out, standing like a knight-errant upon his 'antiquas vias.'

" I had forgot to tell you that Lord Northampton also
spoke to a point of conciliation; in fact, there was too much
of this, for I sat next to De la Beche, never lost my temper
for an instant, asked him to dine with me, and all ended
a l'aimable,' and would have done so without any of the
surpassing efforts of these ' good Samaritans.'

" Buckland was particularly happy in assisting to demolish ' Greywacke' by pulling old Greenough up, who with
himself had declared a mass of rock in the Alps to be good
' *grauwacke*,' which proved to be full of Tertiary shells;
that he had seen very good ' grauwacke' in oolites, in red
sandstone, in coal—in short, in everything, and therefore he
did think with Conybeare that it was ' Jupiter quodcunque
vides,"and agreed with us in the fitness of using it hereafter
entirely as an adjective or expletive.   Q. E. D.

" It was right well that I was *not* absent in Paris, or
things in your absence also might have gone *pro tempore*
against us.—Ever yours,   ROD. I. MURCHISON."

A fortnight later the two Knights of Cambria and Siluria
were ready with their own conjoint paper on their change of
view regarding the geological position of the rocks in Devon
and Cornwall—a change which had afforded one of the
most effective shafts to their opponents in the contest.

In this memoir the term Devonian was proposed as a
substitute for Old Red Sandstone, to include the rocks
lying between the Silurian and Carboniferous systems.[1]
The authors, accepting Lonsdale's suggestion, boldly applied
it not merely to the limestone of South Devon, to which he

[1] The first publication of this proposed new geological subdivision
appears to have been that in the *Phil. Mag.* for April 1839, p. 259.

originally restricted it, but to all the old slaty rocks of both Devon and Cornwall, and even expressed an anticipation that it might be found capable of application on the Continent. To quote their own words, adopted by Dr. Buckland, this was " undoubtedly the greatest change which had ever been attempted at one time in the classification of British rocks."[1]

It was, without question, a most important change in geological nomenclature, and before long it met with recognition and adoption all over the world, insomuch that the term "Devonian" came to be as familiar a term as Silurian or Cambrian had become. And yet we must admit that, though exceedingly ingenious, it was based rather on what seemed probable than what had been proved to be the case. Had the authors simply declared that their Devonian rocks occupied a place somewhere between the base of the Coal-measures, or upper part of the Mountain-Limestone and the Silurian system, their position would have been unassailable. Their identification, however, of the Devonian slates, limestones, and sandstones, as the true equivalents of the Old Red Sandstone of other regions, left out of sight the fact that a great thickness of Lower Carboniferous rocks was on this view unrepresented in the south-western counties, and hence that a portion at least of their Devonian series might really be Carboniferous. Many years afterwards, as will be told in a later chapter, this now obvious objection was started and argued with great vigour and cogency by the late Mr. Jukes.

So greatly have the rocks in Devon and Cornwall been disturbed since their formation, that even now, though they

---

[1] *Trans. Geol. Soc.*, 2d Ser., v. 691, and *Proc.*, iii. 226.

have been examined over and over again by geologists without number, considerable dispute is still held over their true structure. In their memoir to the Geological Society, Sedgwick and Murchison indicated that what they had made out among these cleaved and fractured rocks might not improbably explain some parts of Continental geology, and there was likewise the probability of new light being obtained from the foreign rocks to clear up the obscurities still remaining at home. They had even stated their intention of personally seeking information on these points. Murchison began to think of putting this proposal into practice, and talked at one time of Scandinavia, at another of Belgium, or of the south of Ireland, and again of the Eifel and Westphalia, as the proper ground to begin upon. He urged his colleague to make the tour a conjoint one, and pressed upon him the needfulness of trying to break loose from the ties which seemed to bind him too closely to Cambridge or the Chapter of Norwich.   Thus, early in the spring he wrote, " I was glad to see your handwriting, albeit you wrote in a state of exhaustion.   Allow me, as your true friend, to urge you to make more than an *ordinary* effort without delay to shake off the Norvichian trammels to such an extent as will enable you to do that *something* more in field-geology without which your labours are incomplete and your general views cannot be established.   You say you are junior in the Chapter, but surely you can contrive to get off for one year a month earlier than usual . . . . Pray, therefore, make your arrangements so that you WILL take your fling ' coûte que coûte.' "

Three weeks later his feelings were expressed to the same correspondent as follows :—" I am so sick of the town, and so oppressed with the feeling that I ought to be *at work*, that somewhere I will go in the middle of May. I may, however, defer my Scandinavian tour if I can meet with no playfellow; for in those cold and dreary wilds a solitary tour is out of the question. Belgium, the Ardennes, the Eifel, Taunus, and Harz may be a substitute, and most of this I can work away in until you join me, for I gather from your letter that some portion of this country is your aim. I must be at Birmingham, but I shall make it a stepping-stone to Ireland, where I shall remain till the rains drive me out. Thus we may unite at points of essential interest."

On the 7th April, having meanwhile changed his plans again and again, he wrote once more to Sedgwick about the foreign tour, thus:—"Your letter reached me at Christ Church before we left the Bucklands yesterday, where we passed three pleasant days. . . . . I stuck like wax to B. to get knowledge from him about Normandy and Brittany, and ended by carrying off his maps and two or three sheets of memoranda. . . . . You call me a weathercock, and so I am, but, I hope, for the only object about which I occupy myself in the world. My plan is now definitively arranged. On the 1st May or a few days after, start for Antwerp and Liége—floor that tract in a week with Dumont and D'Omalius and Buckland's section; traverse by Spa and make a round to Treves, perhaps taking a peep at the west side of the Eifel and back to Paris—ten days there before any of the savans have left it, fill myself with knowledge and buy all maps, etc.; down straight to Caen, and there meet Adam Sedgwick in first week of June at latest, and commence work forthwith by the coasts of Normandy amid the Silurians. In two months we shall have *gutted*

everything, and bagged as many ' chouans ' in La Vendée as
we please.   It would be quite useless for you to go to Paris
and lose time.   I will get the lesson for us, and we shall do
the trick quickly ; back to Birmingham for the 26th, and
in the first week of September over to Ireland, where C.
Hamilton and Griffiths will throw us in three weeks into
every good cover, and we shall be home again for October
shooting."

In spite, however, of the minute detail of this " definitively arranged" plan, it was in the course of a week or two
completely changed.   The-final arrangement settled that the
two old friends and fellow-labourers should once more wield
their hammers together on the banks of the Rhine.   The chief
point to be ascertained was whether or not there existed on
the Continent a series of rocks having a peculiar assemblage
of fossils, and passing upwards into the base of the Carboniferous and downwards into the top of the Silurian rocks.
If such a series could be found it would amply justify the
Devonian nomenclature.

Murchison started first.   Taking Paris on his way, he
there attended a meeting of the French Geological Society,
of which he had now become a member, and had a fight
with some of his scientific friends over the claims of the
so-called Devonian rocks to the dignity of being styled a
" system."   He stuck to his point, however, here as well
as elsewhere, and, notwithstanding objections and protests,
both at home and abroad, succeeded in establishing it in the
general geological literature of his time.

The halt at Paris was brief.   Before the end of May the
work had been begun in the heart of Rhineland.   From
Trèves, Murchison wrote to his wife :—" ' *In fine respiro*,' as

I said to myself whilst I walked up yesterday under the fine beech-trees from the little frontier station, and found myself in Prussian land, fairly free of the 'Grande Nation' and all its lies, *émeutes*, and bombast. Thank God I am now in a country I like (people and landscape, with geology of all sorts in the fore and background). I blessed the first glimpse of the vine-tending nymphs, with their Swiss-like broad-topped white caps, and the men with their round slouch-hats, honest German faces, and great jack-boots. Thenceforward all was changed for the better—capital macadamized roads everywhere, postilions with horns; the Prussian arms and eagle marking discipline, order, and comfort everywhere.

" I leave to-morrow morning in a little carriage which I hire (I shall buy one at Frankfort, where they are excellent) passing to Bingen on the Rhine, by Oberstein and Kreuznach to Frankfort. I am here in Cambrian and Longmynd rocks, with overlying red sandstone and muschelkalk. *Portez vous bien.* I wish you were with me, and that we had to pass three or four months quietly in this delightful country, to which I hope indeed we may return, for I shall have plenty to do another year."

From Frankfort on 2d June he informs Mrs. Murchison, " I have bought a Vienna carriage, and a very nice one, which I hope will please the Professor. Finding by his letter of this day that he does not leave London till the 12th, I had almost resolved running away to the Fichtelgebirge to see Count Münster and his collections, and to make a section of that chain, where I believe there is much Devonian; but second thoughts have convinced me that it is better to do one thing well than two things badly. So I stick to the right bank of the Rhine, the contents of which I hope to sweep out so as to fill two portmanteaux (now empty among my carriage boxes), and send them off to Lonsdale's care before the Professor meets me at Bonn."

Meschede on the Ruhr, June 1839.—" Having finished my ' Abendessen,' consisting of a fresh trout, some asparagus, and eggs, I am now smoking my pipe in a very neat clean room overshadowed with trees in this little town of the Lower Rhine, which doubtless you never heard of before. This morning I came hither by Alpe and Bolstein. I have now gone clean across the region, and have looked into the zoological and mineralogical contents of each zone of rocks, as well as their geological relations. What I have to say will surprise you. I do not believe there is a Silurian bed among them, and I am more than disposed to think that the whole is Devonian, except, perhaps, the westward flanks. There are no Eifel fossils here. The limestones are undistinguishable from those of Plymouth and North Devon, and the organic remains are all of the same classes which occur in those rocks—*Goniatites,* large *Spirifers,* etc. To a person bothering and losing himself in details, the geometry of the country is puzzling, as the same zones are repeated several times, both on the north-west and south-east side of the axis. To-morrow I march upon Arnsberg, and thence into the Düsseldorf coal-field. If my conjectures are right, I shall find there Devonian passing conformably under it, and I shall then retraverse to Cologne and Bonn, and prove the case again by other sections. So that, when Sedgwick joins me, I flatter myself that part of the campaign (and which I always thought would be the key to the whole thing) will be in my pocket, and I shall have swept the right bank of the Rhine. So much for unfortunate Grauwacke and all its *Kieselschiefer* and *Dachschiefer*, in the midst of which I am writing. . . . You need not boast too much of my geological hits, as some of them may fail."

The caution in the last sentence of this extract was not unneeded. For the writer had evidently determined to do as clever a piece of geological strategy as he could before his equal in command should join, and he was naturally desirous to make his sections bear out the interpretation which they first suggested to him. But he had already gone wrong in some of his notes, and further errors and corrections were in store for him.

After about a fortnight of such marching and countermarching in search of a good base-line of operations for further conjoint movements, he was joined by the Professor. We resume the extracts from the letters to Mrs. Murchison.

Bonn, 15th June.—" If I have my own way I shall not go near France again this season, at least not till the autumn, and after Birmingham.[1] The mine I have opened here is well worth all our time and attention, particularly when coupled with the Harz and the other ' transition ' tracts of N.-W. Germany.

" As I was sitting under the linden-tree with Oeyenhausen and his lady, not forgetting old Nöggerath, up walked the Professor, and after drinking several jorums of ' Maitrank,' he is now gone to bed. He is delighted with what I have done. I have already convinced him that our whole summer's work will and must be in Germany. We have a grand field before us, and I have already provided a certain

---

[1] The British Association Meeting of 1839 at Birmingham.

key.   In this case I shall return by Belgium in the middle
of August, and after settling Birmingham and our household
affairs, may make a run of three or four weeks to settle the
French affair, which is in a nutshell."

Göttingen, 24th June.—" Since I wrote to you at Bonn
only a week ago, we have done stout work, and travelled
over much ground.   I took Sedgwick back to my key, and
satisfied him of all the main points, which are, indeed, as
clear as noonday, and we have since been puzzling out some
minor difficulties, with which we shall have to contend when
we revisit the region of the Rhine. . . . A most capital
table-d'hôte seems to have put the Professor into working
order.   I hope, therefore, that in a few days we shall hear
no more of his dyspeptic symptoms, which far exceed in
variety any which I ever troubled you with.   He is, however, in very good spirits, and we get on famously.   I have
become very rubicund and jolly, as I always do on work,
with hands as brown as a gipsy's."

Ballenstädt, 1st July.—" We have, thank our stars! nearly
cleared the Harz ; and, though the weather has been of the
most oscillating nature, with severe frowns, we have had
some charming smiles, which enabled us to do our work and
peep into three of the most lovely valleys—the Lauterthal
near the western end of the chain, the Okkerthal near Goslar,
and the Bodethal, about ten miles west of this place. . . .
Sedgwick is as well as I ever knew him,—eats, drinks, and
digests like a Hercules, and is. in great force.   Indeed, we
are both quite well, though the weather is most untoward,
and fresh storms are gathering around.   The geology of the
Harz is very interesting, but complicated. . . . We sleep
in a fresh Principality daily.   All the kings and dukes of

Germany seem to have slices of the Harz, and their respective strips of land run towards the Brocken, like the spokes to the box of a wheel."

Frankfort, 15th July.—" We have now done the Fichtel-gebirge ; and as we travelled here almost without stopping I have been my own bagman. Count Münster was all attention, and his museums delighted us. The Upper Franconian geology was not quite so good as might have been ; but we did all that could have been done. The rocks are two-thirds Devonian, and some Carboniferous—no Silurian."

While these labours were in progress in Germany, other transactions, involving a good deal of Murchison's future comfort, were going on in London. Mrs. Murchison, with the full sanction of her husband, was negotiating for the sale of their house, now in Eccleston Street, and for the purchase of the well-known Belgrave Square mansion, in which he spent the last thirty-two years of his life, and which in his occupancy of it formed one of the hospitable scientific centres of London. This purchase is alluded to in the next letter of the series.

Ems, Coblenz, 27th July.—" The furnishing of our *grande maison* may be done so leisurely as not to fatigue you, and I trust we shall be there for the rest of our lives. At all events, you will have a good airy palace to live in, even should I prefer this tramping life, which I am destined to lead for the few years of bodily activity which remain for me, should I survive to middle age.

" Our last traverse to and fro through the Nassau country has answered in some respects. We were both highly delighted with the work on both banks of the Rhine, between Bingen and Coblenz, which we performed in boats, carriages, and on foot, disdaining all the smoking steamers. Here we are for the day, in this most picturesque watering-place—by far the prettiest of all the Rhenish baths, and doubly interesting to me, because here the first true Silurian rocks which I have seen in any part of Germany on the further bank of the Rhine are in great force—fine scarps and lots of fossils."

Deutz, 31*st July.*—" We have made our last round in the Westphalian region and the right bank of the Rhine, and we are now on our way into the Eifel, in which, after certain zigzags, we shall reach Trèves. I have little worthy to communicate except on geological subjects, and on these little new. In fact, I am quite tired of this bank of the Rhine, and am most anxious to break ground on the opposite. The only thing which annoys me in my work is, that although we have got excellent descending sections from the coal-measures to the bottom of the Devonian or Old Red system, into which *all* the greywacke of the right bank of the Rhine falls, still not a trace can I obtain of Ludlow, though the Wenlock appears on points, and thus we want the connexion which exists in England. It is this which we are to find in the Eifel and the Ardennes. . . . I am swollen out like a German, with hands as brown as tanned leather."

As one of the General Secretaries of the British Association, Murchison required to be present at the meeting, which this year had been fixed for Birmingham. Very unwillingly he quitted the field-work on the Continent and hurried to London. Before joining his colleague in the Secretariate, Professor Phillips, he found time to send him a brief report of his doings with Sedgwick.

London, 18*th August.*—" I arrived last night from Liége, in thirty hours, having left Sedgwick on the Meuse, in full

cry with D'Omalius and Dumont.   I am happy to tell you
that the Devonian system now rests on a basis quite
unmoveable, and that the coal-field of Devon will after this
promulgation of our new data, never more be contested.
Even the sturdy Williams will be swept away !   It was the
observance of the leading facts of the case during my first
month's work, which led me to form a decided opinion that
Sedgwick and myself ought to give up one whole summer to
the establishment of our views, by devoting ourselves entirely
to the Rhenish Provinces and Germany ; and no sooner did
he see the outlines of the case than we resolved to abandon
Brittany, at all events till the autumn, and to stick more to
the classic regions of our science, in which as yet the alphabet
of the oldest strata remained to be pointed out.   To the
Rhenish Provinces we have added the Harz and the Fichtelgebirge, and I return, after having travelled the better part
of 3000 miles, and satisfied with the results."

Next day, full of his new work, he could not refrain from
introducing it thus in a note to his friend Whewell :—" To
tell you of all the wonderful exploits of the Cambrian and
Silurian knights, and how many a dreary rock of grauwacke
they tapped before one of their followers could be found,
must remain for another day.   Grand, however, is the
Devonian field on the Rhine, the Harz, and the Fichtelgebirge.   So you see we have been moving."

The geological doings at the Birmingham meeting of the
British Association proved somewhat tame.   No great
paper made its appearance.   Perhaps the most important
communication in Section C was Murchison's own account
of what Sedgwick and he had done on the Rhine and in
Westphalia.   But that account was necessarily incomplete,

and even inaccurate, seeing that the work had not been brought to a close, and the later rambles of the autumn led the two explorers in some respects to modify their earlier conclusions. The attention of geologists had now been seriously awakened to this settlement of the true age and meaning of the " Devonian System." Several other labourers were in the field, and there could now be no doubt that the problems would not be thrown aside until their solution had been found.

A shade of sadness hung over the gathering of the geologists at Birmingham. The day before they met, William Smith died. He had lived to see his work bearing abundant fruit in every corner of the globe, and now, full of years and honours, he left the harvest to be gathered by younger generations.

At the close of the Association meeting Murchison hastened to the Continent again. Before rejoining Sedgwick, however, he went to Boulogne to attend the " Réunion extraordinaire" of the Geological Society of France, which was held this year in that town. He had instructions from Mrs. Murchison, that while discussing "Devonians" and dinners with his French acquaintance, he should take this opportunity of obtaining some additional furniture for the " airy palace " in Belgrave Square. Here is a part of his report to her:—

Boulogne sur Mer, 12*th Sept.*—" Having been out daily from half-past five till dark, I have had no time for ' furniture' thoughts. It so happens that owing to my having more knowledge of the older rocks than other geologists here, I have been obliged to become a sort of cicerone and orator, and yesterday evening, in the great library, the Mayor of Boulogne and many French present, I delivered myself of an hour of Silurianism, and explained the relation of the old rocks of this country. The effect of my discourse was to destroy the coal-boring mania in rocks of Silurian age. They have a poor little coal-field here which lies low in the Carboniferous Limestone group, and this being immediately recumbent on my Silurian schists and shales, they have (their little upper concerns being about done up) been poking at great expense, and with the money of unfortunate shareholders, into my Stygian abysses. The 'actions' or shares fell 50 per cent. by my speech, and, notwithstanding that I told unpleasant truths, I was warmly applauded.[1] I should have been off to-day, but I was so pressed on all sides to remain that the departure was postponed till to-morrow, when I proceed [with De Verneuil] by Calais."

Bonn, 19th Sept.—" We arrived here yesterday afternoon (*i.e.* M. De Verneuil and myself). I was delighted to find at Spa my little old vehicle, which Sedgwick had left there. As for the chaise seat, he had carried away the key, but on breaking it open we found his *best* coat, some maps and books, and a long well-used and highly-scented tobacco-pipe, all in harmonious keeping.

" We found S. waiting for us, having just returned from an expedition up the Rhine. He is in very good health and spirits, and this afternoon we shall take the field—a valiant triumvirate,—our force being strengthened by De Verneuil's good knowledge of organics of the older rocks ; but whither we shall march, ' Dio lo sa.' I find Sedgwick much bothered and disconcerted about many essential geological points,

[1] From the official report of the Society's meetings, however, it would appear that his views as to the impossibility of finding coal in the older rocks were not unreservedly accepted by his scientific brethren. See *Bulletin de la Soc. Géol.*, tom. x. p. 417.

and much disposed to go into a ' chaotic' state, but I hope we shall put up our horses and come to some clear general conclusions in spite of the apparent hotch-potch of this volcanized country.

" The Walloons are an odd, mongrel people ; the country hideous—high bleak moors, with all the features of the worst parts of the Highlands, and no redeeming grouse. We slept at a great farm-house converted into a sort of caravansary inn.   We had storms and wet in passing through the Eifel."

Lutzerath in the Eifel, 8th Octr.—" I have been a lazy correspondent, but a most active workman.   The days are short, and though up daily at five (by candle-light) we are soon benighted.   Yet, with all, since I wrote we have done a great deal.   From Coblenz we journeyed by the river to Limburg on the Lahn, and thence passed over the Wester-wald, a high basaltic region, to Dillenberg, where we had a famous excursion on foot, headed by a little broad-shouldered clever Prussian bergmeister, who, booted and spurred, led the way (pipe in mouth and hammer in hand), followed by S., De Verneuil, and myself and an English miner.   We got many additional fossils. . . . . At Limburg De Verneuil took leave of us to run through the Eifel quickly to Paris. He is an excellent companion, and of a charming 'temper, never making a difficulty, and a thoroughly gentlemanlike Frenchman ; O how different from a sulky Bull !   Take this for an example :—His travelling equipage, consisting of a *little* leather bag (the size of a shooting bag) was left be-hind at one of our stations.   This was forgotten before we reached our next post, where, caressing a great German pointer, the animal flew at him and bit his lip through.   A

little *eau de vie* cured the wound, and on we travelled, he
as merry as ever, and ready again to play with the next dog
at the next inn.   Arrived at Dillenberg, where the inn is
kept by an old Frenchman with three or four daughters, De
Verneuil was soon at the old piano, delighting the girls with
new versions of all the last Parisian airs (he plays very well),
and in ten minutes the gayest Mademoiselle was in full zest
at a duet with him—one of Strauss's last waltzes.   Without
a shirt, without a razor, without shoes, nothing daunted, he
was up at daybreak, and ready before us for the field,
equipped in one of the old innkeeper's pairs of trousers and
a pair of thick shoes.   Reaching home, his thin boots and
pantaloons were ready.   A village barber shaved him, and
being invited to dine out with the young English miner
and his sisters, De Verneuil completely beat Sedgwick and
myself in his toilet, notwithstanding our trunks and bags.
I was quite sorry to lose him, and I believe he equally
regretted to quit us.   He has been of great use, from his
intimate knowledge of species, and I think we have been of
use to him in geology."

The work was now prolonged into the Eifel, where
further mingled interest and difficulty met the travellers.
The autumn had been making rapid strides towards winter,
as dark mornings and early nights reminded them.   There
were problems in that strange region of ancient slates and
modern volcanoes which they could then find no means of
solving.   Nevertheless they considered that they had
achieved enough for one season.   And so quitting the
grauwacke rocks of the Eifel and the marvellous volcanic
cones which overlie them, they dropt down the Moselle by
small boat, hammering here and there by the way, and send-

ing their carriage by the road. The next letter reports as follows :—

" I now write from the middle of the ' Herzog von Nassau' steamer, floating down the Rhine, and within an hour of Düsseldorf. We had a most charming voyage in our little cock-boat down the Moselle, and reached Coblenz last evening with heads full of grauwacke and lordly castles and dark gorges. To-morrow will see us at work in Westphalia for the last time—our third visit to some spots. We may occupy three, but perhaps only two, days in this work, and then we may sail for England from Rotterdam on the 15th or 16th, and reach London on the 17th."

Soon after their return to England Murchison sent a long account of their autumn campaign to their common friend Phillips. From that letter it is evident enough that the writer did not feel over-confident in some parts of their recent Continental work, and indeed, that in certain main parts his colleague and he were not yet in agreement. But they had still a great series of specimens to be critically examined and compared with those from Devon and Corn-wall. Much of the winter and early spring was given to this task, with the effective and indeed indispensable assist-ance of such friends as Lonsdale, Sowerby, and Phillips. As the boxes were one by one examined, alternate light and darkness passed over the minds of the examiners. At one moment the field-work which seemed to have been so decisively settled by the two geologists began to look doubt-ful, then it grew more than doubtful, then it seemed all right again, and finally it had in part to be discarded as the true reading of the fossils came bit by bit intelligibly out of the examination. Sedgwick remained at Cambridge, but he had from time to time copious bulletins of progress from Belgrave Square. The following extracts may serve as a specimen :—

"*Decr.* 8, 1839.—MY DEAR SEDGWICK,—I have been intending to write to you for some days to keep you *au courant* of the examination of our 'kists' and their contents, and of the views which have been gradually opened out in my mind, and which have now brought me back to the *status ante bellum,* or, in other words, to the same condition of mind, or nearly so, in which I was when you joined me on the Rhine."

[Here follow eleven pages of detail regarding the bearings of the fossil evidence on different parts of their work in Germany.]

"Thank God! I now see daylight again. All our follies proceeded from our attending to these cursed mineralogists and gentlemen who deal in 'symétrie de position,' whose doctrines will now, I bless my stars, go by the board.

"Do not think me crazy, for if this letter is too short to lead you into your former true path, I hope the 'pièces justificatives' [*i.e.* the fossils] which cover my whole rooms will do so.

"What we ought to do is to write a memoir on the right bank of the Rhine, viz., Westphalia and Nassau, with illustrations of similar tracts in the Harz and Ober Frankwald (Fichtelgebirge), and I pledge my life that if plain facts be laid before plain geologists, there will be no escape from my present induction.

"Adieu—once more redivivus, although you had wellnigh killed me."

The result of these laborious deliberations was at last a complete accord on all the main features of the question, and the consequent elaboration of another great paper for the Geological Society.[1]   We get a characteristic picture of Murchison in the following account of these preparations :—

Feb. 25, 1840.—" MY DEAR PHILLIPS,—I thank you for Austen's list, as (if to be depended on) it adds one or two good clenching fossils to a list already too strong to admit of any doubt as to the identity of the uppermost Grauwacke system of the Continent and the 'Devonian' as defined by Sedgwick and myself.[2]   I have arrived at this conclusion for many months, and only waited the coming to town of my colleague to open the campaign.   Now that he has been here, and that we are all agreed, the course is clear, and we shall soon give a grand memoir to show that the uppermost Grauwacke of both banks of the Rhine, as well as the three members of Dumont's *Terrain anthraxifère* (supposed by him to be Silurian), as well as the major part of the Harz and of the Fichtelgebirge, are true Devonian, passing up into Carboniferous strata, and reposing on Silurian. . . , I am now highly delighted in having insisted on the ' Old Red ' as a system, and on my prophecy of what it would turn out in fossils.   I too, however, have made my little mistakes, and I will thank you to allow me to amend some words in my

[1] On the Classification and Distribution of the Older Rocks of North Germany, etc., read 13th and 27th May 1840, and published in vol. vi. of the second series of the Society's Transactions.

[2] This was one of the points on which perfect unanimity was not reached until after the two fellow-travellers returned to this country, Sedgwick having a suspicion that the rocks of Rhineland and Westphalia, which Murchison was inclined to rank as Devonian, were really Upper Silurian. The grounds of this suspicion, and the difficulty of forming a satisfactory conclusion, are well stated in the paper last quoted (*op. cit.*, p. 226).

communication at Birmingham.[1] Again, in returning by Boulogne I gave a field lecture, and, supposing that De Verneuil, Dumont, and others were right in Silurianizing these tracts, I chimed in with the error without looking for fossils.

"I am going to Paris in ten days to read a memoir on the Boulonnais, all the fossils of which have been sent to me, and they clearly Devonianize it. . . . We propose our triple subdivision of Devonian, Silurian, and Cambrian for Europe. Buckland has given currency to our views in his speech, and Greenough has closely imitated our reform of Devon and Cornwall. So at last all is settled as to the great boundaries."

The brief visit to Paris, alluded to in this letter, proved to be a pleasant and by no means unprofitable one. Dinners at the embassy, soirées, evenings at the opera, and other amusements, helped to dilute the draught of science which Murchison had been quaffing so vigorously for so many months. His letters convey a droll jumble of mingled science and festivity. Writing to Mrs. Murchison (April 4), he describes a soirée at Lady Granville's. "There I saw every one," he says, "not excluding Thiers, to whom I was presented, and had some chat. He seemed to be delighted to hear of Guizot's good reception in England, and called him 'un homme éminent.' Thiers is the drollest little body you ever saw, more like Dick Phillips the chemist, with his spectacles, than any one I can recollect at this moment. I heard him to-day in the Chambre des Deputés—a short, clear, and pithy speech, and I can understand how and why he rules.

[1] See *ante*, p. 279.

" To-day I had De Verneuil with me from nine to one, when we adjourned to M. de Meyendorf's, who starts to-night for Petersburg, and with whom we arranged a Russian campaign for June, July, and August. It is agreed (if I do not change my mind) that I sail for Petersburg the 25th May, De Verneuil coming to meet me some days before. The advantages are too great to be lost, both as respects the Russian *factotum* and administrator, and De Verneuil."

Among the hospitalities, he was especially pleased with a soirée or banquet at which he was entertained by a number of the leading geologists of Paris, a dinner from " old Brongniart, in the most hospitable form, with lots of fossils in 'sucreries,' " and a sumptuous entertainment in his honour from M. Élie de Beaumont. In return for these kindnesses he gave a dinner at the " Rocher de Cancale," to a company which included Arago, the two Brongniarts, Élie de Beaumont, Nöggerath of Bonn, D'Orbigny, Valenciennes, Russeger from Egypt, D'Archiac, Boué (then fresh from Turkey), and De Verneuil.

The paper on the Boulonnais was well received at one of the best meetings of the season of the Geological Society of France. Alexander Brongniart was in the chair, and an interesting discussion followed the paper, some of the speakers impugning the right of the Old Red Sandstone to be regarded as a *terrain*, and Murchison standing up stoutly in its defence.

After these few weeks in Paris, passed in this pleasant way, he returned to London, having now but little time to prepare for that Russian campaign, the plan of which he had sketched out. What this plan was, and how it was put in execution, will be told in the succeeding Chapter.

CHAPTER XIV.

A GEOLOGICAL TOUR IN NORTHERN RUSSIA.

AMID the ceaseless revolutions which, during the long lapse of geological time, the surface of our planet has undergone, few tracts have escaped the effects of those movements by which the rocky crust has been crumpled and broken. The older the rocks the longer have they been exposed to these movements, and the greater therefore are the fractures and folds which have been made in their mass. Hence the task of the geologist, though it may be often easy enough among the unaltered deposits of recent times, frequently becomes more and more difficult the higher the antiquity of the rocks which he seeks to interpret. The older the record, the more imperfect and illegible may we expect its pages to be.

It was among some of the older chronicles of the geological record that Sedgwick and Murchison had now been at work for many years. With rare sagacity they had succeeded in eliciting the evidence of the order of succession among some of the oldest and most shattered rocks of Europe. They had developed that order in Britain, and as

far as they were able had traced its continuance among similar rocks on the Continent.   Many a time, however, had they
been thwarted and baffled by the obstacles presented by the
dislocations and contortions of the rocks, insomuch that,
although they felt sure that the general story as they had
interpreted it would be sustained by further investigation,
they could not as confidently defend all their details.

In the course of their work, accounts had reached them
of marvellous regions in the north-east of Europe, to which
the underground movements, so disastrous to the rocks of
the central and western tracts of the Continent, had never
reached—a sort of geological elysium, where no volcanoes
had ever broken out, where no "convulsions of nature"
seemed ever to have disturbed the crust of the earth, from
very early geological times; where the most ancient rocks,
elsewhere heaved up into hard crystalline mountains, lay
still in their original half consolidated state, as if the seas
in which they were laid down had only recently been
drained off.   Moreover, they had heard that in these undisturbed rocks fossils were found—shells, corals, fishes, very
like, if not the same as, those which they had disinterred
from Silurian, Devonian, or Carboniferous formations at
home.   Murchison heard still more about these wonders
during the visit to Paris referred to in the previous Chapter.
Evidently some good work was to be done in that Russian
territory.   He might be able among such undisturbed rocks
to demonstrate by a new mass of evidence the order of
sequence already determined in Britain, and to show that
instead of being a mere local arrangement, that order was
really the normal one for Europe, if not for the whole globe.

With De Verneuil as his companion, the journey would

probably at least be an enjoyable one, and that naturalist's great knowledge of fossils would be of inestimable service. The plan was accordingly sketched out, and forthwith put into execution.

The two fellow-travellers started in May from London, and with no important halt journeyed straight to Berlin. It was through the German geologists, and notably from Humboldt and Von Buch,[1] that Murchison had learned what he

[1] Murchison's obligations to Von Buch are well shown in the subjoined characteristic letter, which further illustrates the estimation in which the Silurian work of the English geologist was held by the highest geological authority of Germany :—

"BERLIN, 23 *Février* 1840.

"Il est certain, Monsieur, qu'il est facile d'être savant, et même très savant, quand on tient une clef en main, comme votre superbe ouvrage. . . . . Nous serons donc Velches, et les noms de Llandeilo flags et de Caradoc nous deviendront tout-à-fait familiers, quoiqu'ils se ressentent un peu de leur origine montagnarde. Je tache à les appliquer aux diverses couches de l'Allemagne, avant même que vos savantes et laborieuses recherches de l'année passée nous auront dévoilées les secrets des montagnes germaniques ; et certes, il faudrait être sans intérêt si on ne croyait voir quelque lumière, votre ouvrage à la main. Mais, comme une huître d'un banc d'Angleterre n'est pas une huître du Holstein ou d'Italie, quoique de la même espèce, de même j'ai un peu d'appréhension que l'Allemagne quoique se plaçant dans le même ordre que vous avez si savamment établi, pourrait facilement ajouter quelque nom barbare à votre série des couches, et au contraire voir s'évanouir ou Wenlock Shale, ou Llandeilo, ou quelqu'autre couche très bien caractérisée. Chaque pays porte un caractère à soi, et de vouloir faire entrer des couches qui sont caractérisées par des productions qu'on ne retrouve pas dans un autre système de montagnes, de vouloir les faire entrer dans une case de la série établie me parait vouloir l'étendre dans un lit de Procruste . . . .

"Vos belles figures m'occupent sans cesse, et le vol. 2 de votre bible géologique ne sort presque pas de mes mains. Avec quelle satisfaction ne doit on pas voir que vous avez vous même éclairé la partie difficile des trilobites ! Plût au Ciel, que d'autres géologues voullussent suivre un si bel exemple, et ne pas abandonner la détermination des espèces aux naturalistes de cabinet, qui ne peuvent pas étudier les différentes modifications des êtres organiques, qu'on observe en place, et qui érigent en espèce chaque individu qu'on les présente. . . . .

"J'avais cru, avant la publication de votre ouvrage, que ces couches du Nord pourraient bien entrer dans le système Cambrien,—je vois depuis

knew of Russian geology. Hence he made for the Prussian capital, with the view of gathering together as full notes as possible of all that was then known on the subject. Among

votre envoi que le caractère Silurien y est encore décidemment prononcé, par les Orthis et par les coraux, depuis le Ludlow jusqu'au Caradoc ; mais Dieu me garde d'y vouloir reconnaître un Caradoc limestone, un Caradoc sandstone, un Caradoc shale. Le cortège de ces Princes doit changer d'après les localités, et le voisinage des Diorites, des Hyperites, des Granits donne un aspect bien différent aux couches subordonnées, que n'ont les couches d'argile et de sable de St. Petersbourg. . . .

"Le superbe Holoptychius Nobilissimus et les planches qui suivent nous donnent tout-à-coup l'explication de tant d'écailles, qu'on a même voulu adapter à des Mammifères, et elles nous prouvent qu'en Livonie le Système Dévonien est très développé aux environs de Dorpat et de là vers l'Est, jusqu'au centre des collines de Waldai près de Novgorod. Ces couches du Nord s'arrangent à peu près ainsi.

Formation jurassique moyenne. Kelloway rock, Oxford clay, à Popilani sur la Windau, à l'est de Liban. lat. 56½°.	C'est le point *le plus boréal en* Europe ou on connaisse cette formation ; elle est répandue sur toute la partie méridionale des pays Baltiques, même aux environs de Berlin. Ammonites Jason, pollux, polygyratus, Pecten fibrosus la caractérisent : Gryphea dilatata. Les couches supérieures manquent toujours.
I.—Système Carbonifère. Une grande partie des collines Waldai depuis Novgorod jusqu'à Wolotschosk et le long du Wolkov. Le flanc de l'Oural en Asie autour de Bogoslavsk 59¼°.	Productus comoides, punctatus, antiquatus, Mya sulcata, Melania rugifera, Spirifer trapezoidalis, etc.—point d'Orthis.
II.—Système Dévonien. Grès de Dorpat à écailles d'Holoptychius et Calcaire avec Terebratula Livonica, décrite et figuré par moi ; d'immenses masses de Favosites ou Chœtetes capillaris.	Le lac Peipus en est entouré. Les grands champignons de Chœtetes se retrouvent jusqu'à Moscou, pesants des quintaux entiers.
III.—Système Silurien. Couches des collines de St. Petersbourg, Selo, Paulowsk, Pulcowa, Esthonie, falaises de Reval.	Deux Trilobites en abondance. Je ne les trouve pas en Angleterre. Des Orthis en foule, je les ai décrits, surtout Orthis Panderi, Orthis Pronites on cenomala, adscandens ; Orthis elegantula, qui est bien votre canalis ; Orthis radiata, etc. . . .

"Continuez, je vous prie, de nous éclairer et de nous instruire, et comptez sur la reconnaissance de tous ceux qui prennent quelqu'intérêt au globe qu'ils habitent, et surtout sur celle de votre très dévoué serviteur,

"LÉOPOLD DE BUCH."

the friends who lent their assistance were Humboldt, Ehrenberg, Gustav Rose, Von Dechen, and others. In writing to Mrs. Murchison, he thus describes some of the interviews in Berlin :—

" The morning with Ehrenberg was arranged by Humboldt, who accompanied us, and I never in my life enjoyed two or three hours more intensely. To have the wonders of the infusorial creation clearly explained by the discoverer himself, and the whole illuminated by the flashes, episodes, and general views of ' Der Humboldt,' was enough to stir up every sympathy of a naturalist. We little know, at least we do not know enough, in England of Ehrenberg's immense knowledge. He is not merely a microscopic but a great philosophic observer. Humboldt places him in a rank above Cuvier, on account of the superior soundness and accuracy of his discoveries. . . . Tell Sedgwick that I am super-saturated with proofs of the correctness of our views, and that I shall be certain to bring home much grist to our common mill."

The following letter gives some further details, and starts a project which, though proposed so long ago, has never been put in practice—an international congress of men of all sciences, superseding for a year the usual meetings of such national gatherings as our own British Association :—

" BERLIN, 28*th May* 1840.

" MY DEAR WHEWELL,—Accept a few lines from your wandering friend. We were too late to catch the Lubeck steamer, so we consoled ourselves with Berlin, where we have been for the last three days resting in intellectual and physical enjoyment with Humboldt, Von Buch, Von Dechen, G. Rose, Ehrenberg, etc. I have seen and learnt much, and have been so fêted, as ' the Silurian monarch,' that it might well turn the head of any one but an old soldier, who knows very well how to receive a *feu-de-joie.*

" The immediate object of my writing to you is that I have been your trumpeter, in my best fashion, I hope, with an ' éloquence vraiment britannique,' in announcing your forthcoming great work, particularly at a great *déjeûner* given to us this morning by Humboldt. I ventured to mention of what great use your book would be to him before he launched his ' Cosmos,' and I hope you will send him one of your first copies, through his relative Baron Bülow. He expressed great regret at never having made your acquaintance, which feeling I augmented by telling him you were the English Humboldt.

" I have long had a project in my mind, which I now intend to broach, and have indeed done so here. Seeing that our various national associations prevent the men of all parts of Europe from meeting each other, I propose that two years hence, that is, for 1842, each nation should abstain for a year to have its local meeting, and that we should all congregate in a central town of Europe. Frankfort, the seat of the Germanic Diet, easily accessible from England, France, and Italy, appears to me the best spot, and that we should honour the close of Humboldt's life by placing him in our chair. No one is so generally beloved, and no one was ever his enemy, and he would give us a fine broad philosophic discourse. If I can [induce] you and one or two strong men to get up the steam, I am sure it would be a really good thing, and productive of much real advancement and enjoyment. Write to me, Pension Anglaise, St. Petersburg, and say what you think of it. I am certain that the British

Association would rejoice to have a year of *relâche* after Manchester, or wherever we may go to next year, and by having so much time to prepare we could make out an excellent bill of fare."

With introductions to the authorities at St. Petersburg, the travellers found their way smoothed for them. To continue our quotations :—" The chief of the douaniers asked for ' Murchison,' and we had the advantage of having our things passed and sealed up with the Imperial arms, so that I might have smuggled a mammoth." Similar good fortune, by the friendly aid of the Russian authorities, awaited them during the whole of their tour in the dominions of the Czar.

After some preliminary sight-seeing, their plan of work was arranged, and all preparations completed. Baron A. von Meyendorf was about to start on a tour through the country to inquire into the state of manufactures and trade in the internal governments. With the view of adding to the value of his report, he induced Murchison and De Verneuil to accompany him, together with Count A. von Keyserling and Professor Blasius. The Baron's objects, however, were so different from those of his fellow-travellers, and his rate of progress through the country so utterly incompatible with adequate geological observation, that after a few weeks' trial they had to part company with him. While he rushed forward to complete his statistics, Murchison and De Verneuil, accompanied by Koksharoff, a young Russian officer, who has since done excellent service to Russian geology and mineralogy, followed at a more leisurely but still by no means a slow pace. For about two months they continued on the move. Passing northwards by the great lakes, they reached Archangel, and made some explorations along the shores of the White Sea. Ascending the Dwina, they penetrated into the heart of the government of Vologda, and sweeping westwards by Nijnii Novgorod, and the valley of the Volga, reached Moscow, whence they returned by the Valdai Hills to St. Petersburg.

The mode of travelling differed very greatly from any with which Murchison's previous geological rambles had made him acquainted. Mounted on a light calèche, some-times with five or six horses harnessed to it, he rushed through the country, over sand, boulders, and bogs, at the rate of often as much as ten or twelve miles an hour. " With four ardent little steeds in hand, all abreast at the wheel, and two before, conducted by a breechless boy who is threatened with death if his horse backs or falls, your bearded Jehu rattles down a slope at a headlong pace, and whirling you over a broken wooden bridge with the noise of thunder, he charges the opposite bank in singing " Go along, my little beauties—fly on, from mount to mount, from vale to vale,— 'tis you that pull the *silver* gentleman—(their delicate mode of suggesting a good tip) ; 'tis you, my dears, shall have fine pastures," the whole accompanied by grand gyrations of a solid thong, which ever and anon falls like lead upon the ribs of the wheelers, followed by screeches which would stagger a band of Cherokees." [1]

It is true that for many a long league such rapid loco-motion by no means interfered with geological observation, the ground being so thickly covered with clay or sand that none of the underlying rocks appeared at the surface. These monotonous tracts deserved the description which

[1] *Quarterly Review*, vol. lxvii. (1841), p. 360,—an article by Murchison on the Russian provinces, with excerpts from his own reminiscences of this first journey in that Empire.

Sydney Smith had once given to him of Holland,—"the place of eternal punishment for geologists, all mud, and not a stone to be found."

Over wide districts of territory there were no inns. The travellers quartered themselves for the night on some priest or peasant, sleeping generally on their own "shake-downs" upon the floor. Nevertheless, they seem to have escaped the "creeping and biting horrors" by which such a berth is usually accompanied. The food being necessarily often indifferent, at every available place they laid in a new stock of provisions, among which roast-beef would appear to have usually had a place. At one wretched village, for instance, it is noted that "we dined on our portable soup, with an egg or two, followed by the inside of our roast beef, the exterior being by this time (therm. 80°) in a greenish, mouldy state." In the towns, however, thanks to the semi-official character of their journey, better fare and more comfortable quarters were secured to them. Thus at Archangel, the governor, together with the English and French consuls, afforded them much help. "Everything," says Murchison, "was light and easy, except two great dinners of twenty-five persons each, which we ate in company of Russians, Germans, Norwegians, French, and English,—all these languages going a good pace throughout the meals."

One of the pleasantest parts of the journey seems to have been the luxury of tea-drinking, especially after days of long, hot, and dusty travel. To sit in a "traktir" and sip tea "of infinitely finer aroma than the Celestial Emperor will ever permit to approach the depots of Canton," or in some forlorn village to set his portable urn agoing, and "at once command a cup of delicious tea," afforded our traveller a pleasure of which the very remembrance continued to have a pleasant aroma about it. We can well imagine, therefore, with what appreciative interest, on getting at length to the great traktir at Moscow, he must " have counted seventy neat waiting-men ready to hand you a cup or a chibouk, and 200 *teapots* arranged in one of the great vestibules of those spacious saloons ! "[1]

The journals and letters written during the tour give a detailed enumeration of the stages, with copious notices of the geology. The writer seems to have been too busy with the rocks to have had much leisure to observe, or at least to describe, what had not a distinct geological bearing. Now and then, indeed, he does make a note of some social custom or other non-scientific fact. Thus, at one of the villages through which he travelled there had been an epidemic among the horses, and the ceremony of blessing the animals was going on as he passed. "A parish priest in his robes was chanting in the centre of a group of horses, whose heads were held around him by various men and women. We stopped the carriage for an instant to see the ceremony. After a short prayer (his books lying before him upon a table) the priest dipped a sort of brush into a bowl of water which he had consecrated, and turning to each horse dashed some water in its face, and afterwards on its flanks. The running back and movement of the horses, the solemn faces of the peasants, and of their wives and daughters, who stood aloft on the high steps and balconies of the cottages, produced a very pleasing subject for the artist, and I regretted for the hundredth time that I had not a

[1] *Quart. Review,* loc. cit. p. 365.

travelling friend who could sketch the scenes of life in this original country."

For the Russian peasantry he conceived a high admiration, which subsequent travelling in the country only confirmed. Their patience, good-nature, courtesy, readiness of resource, and cheerfulness, called forth his frequent praises; nor less was he satisfied with the intelligence and civility of the officials with whom he came in contact. He entered the empire willing to be pleased, and he left it with an almost enthusiastic appreciation which lasted to the end of his life.

Long leagues of jolting over rough roads and byways tried at once the patience of the travellers and the timber of their carriage. Here is an account of their triumphant entry again into the capital :—" Our near fore-wheel, which had been for some time very rickety, fell to pieces as we approached Ijora, so this gave the blacksmith a three hours' job, whilst we were in a horrid hostelry. Travelling on at night, we broke down again within a hundred yards of the post at the gate of Petersburg, and were obliged to sleep here. The wheel renovated, we started, and it again became dismembered five hundred yards from the starting-place. I write this among the Vulcans, doubting if we reach the capital to-day. . . . At length we reached Mrs. Wilson's, on our tottering wheels, on Tuesday the 25th August at 8 A.M."

Murchison was fond of rapid geological work. With his faculty of quickly seizing the salient features of the geological structure of a country, he liked well to move swiftly from point to point, eye and note-book busy all the while noting and recording each point as he went along. During

this first Russian tour there was ample scope for the gratification of this taste.   The general structure of the regions
visited was simple enough, so that a few traverses and the
examination of sections at comparatively few points, gave
the order and arrangement of the rocks over vast areas of
territory.   The difficulties of the task are thus summarized
by him :—

" Three causes impede geological researches in Northern
Russia : 1st, The flatness and unbroken surface of the country ; 2d, The thick cover of drift and alluvium; and, 3dly,
More than anything, the suspicion of the peasants, who never
would give information, inasmuch as they believe that you
are in search of something by which they may be taxed or
oppressed by some order of the Government, or its employés."

And yet, notwithstanding these scruples, a vast deal of
cross-questioning of the natives went on all through the
journey, sometimes not without good effect; for, in their
necessarily rapid traverse of the country, the travellers,
having no guide-book literature to help them, trusted to the
natives for information as to sections worthy of visit on
their route.   At Usting they met the man who had made
the now well-known deep sinkings in the frozen soil of
Yakutsk, in Siberia.   Murchison notes, that after a long in-
terrogatory, he learnt that, with the exception of about 60
feet of alluvium, the shaft to the depth of 350 feet was sunk
in hard grey limestone, with partings of shale and coal.

By taking advantage of all available information, and
making good use of their eyes along the line of journey,
the travellers succeeded, in spite of the flatness, and the
interminable sand, clay, and boulders, in establishing the
order of the palæozoic formations over a great part of

Northern Russia. From a lower mass of ancient crystalline rocks they had made out a most complete and interesting ascending series of Silurian, Old Red Sandstone, and
Carboniferous deposits, not hardened, broken, and crumpled
like the corresponding rocks in Britain, but flat, and only
partially consolidated. So young indeed did these truly
ancient deposits appear, that it was difficult to realize that
soft blue clays and loose friable limestones were the geological equivalents of hard fractured slates and marbles in
Western Europe. Only by recognising in them the charac-

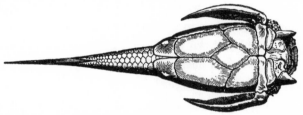

Pterichthys, a Fossil Fish of the Old Red Sandstone of Scotland and Russia.

teristic fossils of the typical districts could their true geological horizon be ascertained.

By much the most important observation which they
made was the discovery of the Old Red Sandstone fishes in
the same beds with true Devonian shells—a discovery the
full import of which will be perceived if we remember the
long and arduous struggle which Sedgwick and Murchison
had had to show that the Devonshire *killas* answered in
point of geological time to the Old Red Sandstone and Conglomerate of other districts. " If I had seen nothing more
than this," Murchison writes, " it would have been a great
triumph for myself and Sedgwick. When we contended that
the limestones and sandstones of Devonshire were of the
same age as the Old Red Sandstone of Scotland, we were met

with this objection, ' Show us a fish of the Old Red in Devon, or a Devonshire shell in the Old Red of Scotland.' Here, then, in Russia I have solved the problem, for these shells and these fishes (species for species) are here unquestionably united in the very same flagstones."

A rapid journey homeward brought our traveller back in time for the meeting of the British Association, which was held in September in Glasgow. The results of the tour in relation to the Devonian question had been so unexpectedly remarkable that he was no doubt anxious to get back to the Association Meeting, where he would have the opportunity of announcing his important discovery. While on board the steamer dropping down the Baltic, he wrote full of glee to Sedgwick, giving an outline of the journey, and of some of the more important geological details. " Our success," he remarked, "has been so great that I am of course in very good humour, which I take the earliest opportunity of communicating to you, hoping that the ' trinitarian' proof[1] which the examination of this vast region has afforded me of the truth of Devonianism will set you up for the winter, drive away all acid and gout, and make you ' Adamus redivivus.'

" Well or ill, I am sure, however, you will rejoice in the splendid and unanswerable confirmation of our views. . . . Think of my audacity ! Here I am without a speech to open the grand congress [at Glasgow], but what I have been scribbling in the steamer. If this finds you in good health, send me a bit of a sky-rocket of a finale, with allusions to Arran, and their coal-fields and their mineral wealth, and their Watt, and their forty-horse-powers, and you will much

[1] He refers to the union in the same strata of the mineral characters and fossils of the Old Red Sandstone with the fossils of the Devonian rocks.

oblige your friend, who, however he may see all these things
floating before him, has not the same power as you of putting them into attractive form.

"I am now more on the move than ever, and having
got the *cacoethes*, I am planning the Ural on one hand, and
the Alleghanies on the other, for nothing short of Continental masses will now suit my palate."

Of the memorably successful meeting of the British
Association in Glasgow in 1840 some notes may be gleaned
from his letters written under the enthusiasm of the time.
Thus, to Dr. Whewell he says—"We never had such good
work as in our geological section; and I am told by Sabine
that Section A was admirably conducted by Forbes. The
opportune arrival of Enke, Agassiz, and Airey gave a great
brilliancy to our last days. From the Duke of Hamilton,
whose palace has been open daily with dinners of fifty persons, down to my hearty friend Thomas Edington, there is
but one feeling of satisfaction. It is, I give you my word, the
only meeting which I have attended where nothing has been
done which I could wish altered, save the statistical display;
all the rest has been done kindly, cordially, and well, which
I very much attribute to the excellent Lord Provost and the
Locals, who have brought together all classes.

"Colquhoun's after-dinner speech—a complete smasher
for the *Times;* the good, manly, unaffected bearing of our
chief [Marquis of Breadalbane]; the very good sense shown
by Lord Greenock; the unbounded joy of my Russian friends,
who kissed me on both cheeks,—all these circumstances, not
omitting the glorious day at Arran, when I lectured to a good
band of workmen, with every peak of Goatfell illumined, and
marched up at the close of the day to Brodick Castle with

the heir of the House of Douglas, preceded by the piper—all these things, I say, have well repaid me for my journey from Nijnii Novgorod, and have more than confirmed the anticipations I entertained of the success of the Glasgow meeting." To that success Murchison himself contributed much. Still holding the office of General Secretary, he had to superintend a vast mass of details which, though separately insignificant enough, combined to determine the success or failure of such a meeting. The kindly, genial President, was not a man of science. Instead of attempting to prepare a scientific address, he very properly left to the General Secretaries the task of drawing up a brief sketch of the progress of science. "It is my fate," wrote Murchison to Whewell, just before the meeting, " to have, in conjunction with Sabine, to prepare a note of the King's speech, to be read at Glasgow." [1]

To this meeting a general interest attaches in the history of British Geology, inasmuch as it brought into notice and into personal acquaintance with the geologists of the day two men who have since made their mark in the literature of British Geology—Hugh Miller and Andrew Crombie Ramsay.

The name of the stone-mason of Cromarty had for some years been known to geologists who took interest in the older rocks as that of a diligent and successful collector of the fossils of the Old Red Sandstone of the north of Scotland.[2] He had recently come to Edinburgh as editor of a newspaper. In the columns of that journal he had begun to publish sketches of the structure of the strange fishes which

---

[1] The project of an international congress of science is publicly proposed in this address. See *Rep. Brit. Assoc.*, vol. for 1840, p. xlvii.

[2] See *ante*, p. 257.

he had disinterred, and graphic pictures of the scenery and geology of the Cromarty coast-line. These contributions had already attracted the notice of some of the leading geologists of the day. Hence a kindly and appreciative welcome greeted Miller's personal entrance into the ranks. The cordiality of his reception was shown by none more than by Murchison, who, indeed, had been largely instrumental in bringing him forward, and to whom he next year gracefully acknowledged his gratitude by dedicating to the author of the " Silurian System " the volume into which the newspaper articles grew—the charming and classic " Old Red Sandstone."

Mr. Ramsay was then a young man, who, betaking himself to Arran, had scoured its glens, hill sides, and shores, and made a large geological map and model of the island. These he exhibited at the British Association meeting, accompanying them with an explanatory paper. His work showed him to possess in so eminent a degree the qualities out of which a good field-geologist is made, that Murchison was greatly impressed with his capacity, and proposed to take him abroad with him in the following year. Though that determination was not carried out, it led directly, as we shall see, to Mr. Ramsay's joining the Geological Survey, and thus opened up for him the path by which he has risen to distinction.

Sedgwick did not appear at this meeting; indeed, he had become so remiss in his attendance at the gatherings of the Association as to suggest that he meant to retire from it altogether. His presence was missed during some of the discussions in the Geological Section, for an observant eye might now have perceived the first speck forming of that dark cloud which, slowly gathering year after year, finally blighted all his close friendship with Murchison, and led him to retire from the society which he had brightened for so many years.   Immediately after the meeting Murchison sent him a letter containing some account of what had been done.   That letter has a special interest in connexion with future events.   It serves too to show the active part its writer took in the management of the Association, as well as his characteristic regard for high social position :—

"WISHAW HOUSE, *Sept.* 26, 1840.

"MY DEAR S.,—Our Glasgow meeting has been altogether the most successful that could have been desired. . . .

"I was compelled to take a strong measure, but one of which I know you will heartily approve, in putting Whewell in nomination as our next President, for the Plymouth meeting.   I say a strong measure, because on my broaching it to him he wrote me a letter of four sides (just before he left us, and in the middle of the meeting) to show that he was in every respect disqualified.   Such, however, was not the opinion of a single person here whom I consulted, and I therefore went on, and he was elected by acclamation, *nem. diss.*   It appeared that the Manchester folks rather wished to have us in 1842 than in 1841, so we were suddenly thrown upon Devon.   To carry out the principles of alternation alluded to in my opening address (which I send you), it was essential to have a man of science at our head. So the staff of science for that meeting are, Whewell, *Pres.;* Snow Harris, Hamilton Smith, and Were Fox, *Secretaries;* and four men of local weight and family to balance them as V.-P.s—Sir C. Lemon, Sir T. Acland, Lord Morley, and Lord Eliot.

"Agassiz's arrival was very opportune, for he confirmed

the identification of the Russian and Scottish fishes. I also resolved to pull out Hugh Miller of Cromarty, and other Scotsmen of the north, and on the last day I gave an *exposé* of all that you and myself did in the beginning of this foray, and held up our sections and our *Dipteri.* Agassiz followed, and ended by naming the curious new winged creature *Pterichthys Milleri.*

"Agassiz gave us a great field-day on Glaciers, and I think we shall end in having a compromise between himself and us of the floating icebergs! I spoke against the general application of his theory.

"Mr. Bowman's memoir contained some good details.[1] . . . . I explained that the outline between Cambrian and Silurian in that region [North Wales], as inserted by yourself in my map, was done without Ordnance maps, and merely to serve as an approximation; that both you and myself were aware of the age of the beds in the Vale of Llangollen, and that some day or other you would roll out what had been for many years in your head and wallet. De la Beche and Phillips pressed me about the natural line of separation between S. and C., on which I replied as in my book, that in many parts a fixed line of demarcation was impossible, but that I was convinced that to whatever extent the same species of fossils as in the Lower Silurian strata descended into your upper group, you could show

[1] The paper referred to here was one in which its author gave the result of some traverses which he had made across the supposed boundary-line between the Cambrian and Silurian tracts of North Wales. He could find no fossils in the so-called Cambrian rocks differing from those of the Lower Silurian series, and stated that "if there be any boundary between the Upper Cambrian and Lower Silurian systems, it must be defined by other evidence than that of fossils."—*Brit. Assoc. Reports,* 1840, Sections, p. 102.

the existence—indeed, that you had already done so both
in Wales and Cumberland—of vast masses of much higher
antiquity which must have a distinguishing name."

After all the scientific and social work of the Association
Meeting at Glasgow had been successfully completed there
began another series of hospitalities.   Not a few of the landed
proprietors, specially those who had taken part in the gathering, invited the more prominent members of the Association
to visit them.   In this way Murchison and his wife found
themselves once more in the heart of the Highlands, enjoying
the scenery and good cheer of that region.   From Lord Breadalbane the General Secretary had some deer-stalking at the
old homely shieling of the Black Mount ; but part of the
journey was planned to include a visit to the north of Scotland, with Agassiz, to look after the Old Red Sandstone and
its fishes.   By the 29th of October he had reached Alnwick
Castle on his homeward journey, whence he writes to Sir
Philip Egerton :—" I believe if I consulted my own happiness
I should do nothing but visit till Christmas, but this must
not be.   Work must be revised, and I have an overwhelming
mass to reduce to order, which if not done before 'the big
wen' begins to fill will never be done.   So I have resolved
even to give my old friends of the North Riding the go-by,
and to stick to the east coast, finishing with Cambridge, and
reaching Somerset House in time for our second meeting in
November.   If you have not been frost-bitten by Buckland
you have at all events had plenty of friction, scratching, and
polishing, before now, and next year you may give us a paper
on the glaciers of Wyvis and the 'moraines' on which you
sport !   I intend to make fight."

To face page 309.

REV. PROFESSOR BUCKLAND, D.D., F.R.S.

Equipped as a "Glacialist," from a sketch by Thos. Sopwith, Esq.

The " frost-biting " referred to the remarkable series of observations by Agassiz among the glaciers of the Alps, and the extension of them to Scotland by Buckland, Lyell, and Agassiz himself. Many years earlier Sir James Hall had directed attention to the way in which the rocks on the surface of the country had been smoothed, polished, and striated, by some great natural agent. He made a careful examination of these " dressed rocks," attributing them to the effects of some powerful débâcles or earthquake-waves, sweeping over the land and hurrying along sand, gravel, and huge loose blocks and boulders. A study of the phenomena of the Swiss valleys, however, had taught Charpentier, and afterwards Agassiz, that the smoothing and scratching of the rocks could have been the work of but one agent—glacier-ice.[1] Profiting by Swiss experience, Buckland had already begun to identify some of Hall's " dressed rocks " and other superficial phenomena, as strictly parallel with those among the Alpine valleys and plains. And now, in the autumn of this year, the great Swiss naturalist, who had come to Scotland chiefly to study Old Red Sandstone fishes, found everywhere, to his amazement, the counterparts of the ice-worn rocks and glacier débris which he had been so intently looking at among his own great mountains. He not merely corroborated Dr. Buckland's identifications, but went so far as to proclaim that Scotland, the north of England, and indeed a great part of the northern hemisphere, had once been actually buried under vast sheets of ice.

So bold and startling a doctrine involved an intimate

---

[1] It is common to attribute the first observation of this geological agency of glaciers to Agassiz. It was recorded by Charpentier, however, apparently as a known fact, five years before Agassiz's observations appeared.—*Annales des Mines*, 1835, viii.

acquaintance with the everyday life and motions of a glacier, which at that time British geologists did not possess. Consequently the views of Agassiz met with little favour. The opposition which Murchison promised them was joined in vigorously by other scientific leaders. Hence fully twenty years had to pass, and a new generation of labourers had to appear upon the scene, before the essential truth of Agassiz's teaching was generally recognised.[1]

But pleasant and useful though this Scotch tour proved to the busy General Secretary, it formed only a kind of interlude in the serious task of interpreting the geological structure of the older rocks of Russia. As he said himself, he had returned from the shores of the White Sea to take his place in the Association at Glasgow. Hence, when once more back amongst his note-books and maps in London, he returned heart and soul to Russian geology. While the incidents of travel remained still fresh in his recollection he wrote the article (already referred to) for the March number of the *Quarterly Review*, on "Tours in the Russian Provinces." While reviewing the works of recent travellers in that part of Europe he reveals, in a characteristic way, his own identity. For there must have been few readers of the gossipy article who did not perceive that its author had been with Moore in Spain and Portugal, that he had subsequently dabbled in art at Rome, that he retained a sentimental affection for the old Highland Jacobites and the doings of those who were "out in the '15," that he was addicted to geological pursuits, that he had spent the preceding summer doing geological work in the north of Russia, and that,

---

[1] See a memoir on the Glacial Drift of Scotland, *Trans. Geol. Soc. Glasgow*, vol. i. Part 2.

in short, he could be no other than Roderick Impey Murchison, though under a somewhat different guise from that in which he was ordinarily known.

The more serious work of this winter appears to have consisted partly in the preparation of the memoir on the continental Devonian rocks with Sedgwick (and, of course, with the repetition of delay at Cambridge and urgent entreaty from London), but mainly in drawing up an account of the Russian journey for the Geological Society. This latter task helped to indicate more clearly the points of defective knowledge which were to be cleared up by the next tour.

That tour had been partly planned before he and his companion, De Verneuil, had left Russia. It was heartily entered into by the Russian authorities, from whom, indeed, Murchison received a flattering request to continue his labours, with the promise of ample assistance. He determined to avail himself of these offers, and strike across the Russian Empire, into the heart of the Ural Mountains. So long and arduous a survey was evidently one which could not be accomplished in a short summer holiday. It would require longer time and more endurance than that of the previous year.

Two Societies claimed and certainly received Murchison's firmest allegiance—the Geological Society, and the British Association. His proposed absence from this country, however, altered considerably his relations to both, and he accordingly made up his mind to resign the post of General Secretary to the British Association. In intimating this design to the President, Dr. Whewell, he could justify his absence this year by the importance of the work he had undertaken abroad, as well as by the fact that he had not hitherto failed to take his share of work at every meeting of the Association since its foundation, and he concluded his letter with the assurance, that when the 29th of July came round, he would not forget the gathering to be held then at Plymouth under Whewell's leadership, but would " drink to their healths if any liquor can be had in the Ural Mountains."

Things had turned out otherwise at the Geological Society, for there, at their anniversary in February, and with the knowledge that he would be absent from England during the greater part of the year, his associates once more placed Murchison in the President's chair, and sent him on his self-imposed travel with all the prestige which such a post of honour carries with it.

As already mentioned, he had formed a wish to help the young geologist who had shown so much geological skill by his model and description of Arran, and that wish had to some extent taken practical shape in a plan to carry Mr. Ramsay abroad with him. The latter, accordingly, came to London about the middle of March; but at the last moment the proposed plan of conjoint travel was changed. This change, at first so bitterly disappointing to his young friend and future colleague, but in the end so fraught with benefit to both, was thus announced by Murchison at the time :—
" Having decided upon going to Russia, and not to America (and I shall be off in ten days), I have unwillingly given up the idea of taking you with me; but, in doing so, I have secured for you a much more lucrative place than any which I could have offered you about myself. Mr. De la Beche has kindly promised to place you on his list of assistants of

the Ordnance Geological Survey.   As the work in question is one for which you are particularly fitted, I hope you will approve of my endeavours to serve you."

Mr. Ramsay has kindly furnished the following reminis-cences of these early days of his intercourse with his future chief:—"I think I must have dined five or six times with Mr. M. during my thirteen days' stay in London; once at the Geological Club, at the Crown and Anchor by Temple Bar, where I first met some of the great geologists whom I had not previously seen in Glasgow at the B. A. meeting.   Mr. M. introduced me specially to old John Taylor, a famous man in the mining world, and much respected and beloved by all the geologists, and indeed by every one.   He was treasurer to the Club.   I sat between him and Major Clerke —an old warrior, with a cork leg, a man of perfectly polished manner, witty, and with a vast fund of anecdotes, some of which were of the complexion called blue.   At that Club meeting, I recollect Sedgwick and Buckland, Phillips, Greenough, Fitton, Lyell, Sopwith, and Owen, and there were others that I forget.   Forbes was then a young man just on the eve of starting to join Graves in the Ægean. The dinner made a great impression on me.   Mr. M., as President of the Society, was in the chair, but I do not recollect anything that took place except the mirth created at our end of the table by Major Clerke and old John Taylor's deep voice and pleasant laugh."   A few days after that dinner the President was on his way to Russia, while his friend joined the Geological Survey at Tenby, there to begin a long and distinguished connexion with that branch of the public service, of which he is now the honoured and esteemed chief.

Two days before starting Murchison sent a parting note to Sedgwick, in which he wrote :—" To cleanse an Augean stable filled with Rhenish, German, and Russian fossils, and to leave the home of the British Association clean swept and all in order, has been no light work for the last fortnight. To make the map for our memoir gave me no small trouble, but now all is done, and the whole concern is ready to go to press, if the Council does not turn crotchety and puzzle-headed. If they do, we must publish elsewhere without loss of time, for the data are good. . . . I am off the day after to-morrow. . . . . God bless you. Go to Plymouth and fight my battles. It is now your turn."

CHAPTER XV.

IT was with a more ambitious programme, and with the advantage of the previous year's experience of the country, that Murchison once more, in the spring of 1841, bent his steps to the Neva. De Verneuil again accompanied him, and shared in the honours and the toils of a still more eventful and successful campaign than any which they had yet undertaken together. The two friends had grown dear to each other. But apart from the ties of mutual esteem, they presented a singularly happy conjunction of qualities for their special scientific work. Murchison's quick eye in detecting the leading elements of geological structure would have been of comparatively minor value without De Verneuil's wide knowledge of the early forms of life, on the determination of which the comparison of the rocks yet unvisited with others already well known was mainly to be based. In their Russian colleague von Keyserling they found an admirable travelling companion, and one to whose judgment and powers of observation the success of their conjoint work in the empire of the Czar was largely indebted.

The route chosen, as before, lay by Paris and Berlin. During a short halt at Paris Murchison had an opportunity of gathering the opinions of the geologists there as to the work which Sedgwick and he had been doing in Devonshire and Rhineland.   He lost no time in letting his friend know the result.   " Every one here," he writes, " is most anxious for the appearance of our memoir, as well as Dumont and the Belgians. . . . . Whatever dubiety may shroud the minds of some of our countrymen, the thing is already quite done as to the Continent.   All the palæ-ontologists are with us, and I am happy to tell you I saw yesterday in Élie de Beaumont's closet the copperplates of the table of colours of the great map of France, in which Devonian, Silurian, and Cambrian are all regularly engraved.

"As you are going to Plymouth this year, I beg you will look about you both inside and outside of the Sec-tion C. . . . . It may be the object of ——— and ——— to mystify our divisions. *But stand to your guns.* The types are clear and distinct, and beds of passage are not to frighten us. . . . It would gratify me much if you could devote an hour to me immediately after the Plymouth meet-ing, and tell me how all went off. . . . . The geological sight here is the Artesian fountain at Grenelle, which I visited yesterday.   It is a noble rush of smoking water— quite a comfortable tepid bath.   Portez-vous bien, my dear friend.   Think of me when I am in Siberia, as I shall think of you holding forth on the Breakwater; and wishing you a happy meeting, and an absence of all gout, believe me," etc.

There would seem to have been only one incident of note in the early part of the journey : Murchison and De Verneuil were all but arrested, in entering the Prussian territories, on the charge of issuing false notes, which they had unwittingly obtained at Paris. They were helped out of the difficulty by Humboldt. Such portions of their short stay at Berlin as could be spared from the hospitalities abundantly offered to them by their German scientific brethren, were devoted to the acquisition of additional information as to what was known of Russian geology. They arrived at St. Petersburg on the 30th of April.

The Russian capital was at that time full of bustle and excitement, on the occasion of the marriage of the eldest son and heir of the Emperor Nicholas. A magnificent series of fêtes had been organized to celebrate the event. Our geologists had determined to see these sights before beginning their work. Besides, Murchison looked forward to obtaining considerable official assistance for his survey. He judged it a good stroke of policy to make the acquaintance of as many of the leading ministers and heads of departments as possible. At the British Embassy he met many old acquaintances, and made not a few new ones, obtaining likewise the much-coveted invitation to the Imperial Palace. How these days of festival were spent is best told in his letters to his wife :—

" The last few days have given us pleasant dinners, at Lord Clanricarde's, at the French Ambassador's, at General Tcheffkine's, where we settled our line of march, at the Minister of Finance's, Count Cancrine, and, yesterday, at Prince Butera's. The last was the most sumptuous of all these feeds, many Circassian lacqueys, and mushrooms in every dish. From General Kisseleff, the Minister of the Imperial Domains, I had a history of the successive denudations of the wood of each region of Russia, and how each denudation had proceeded from south to north. Herodotus describes the regions bordering on Turkey, now grassy steppes, as dense forests. This being for centuries the great line of march of Tartars and Easterns towards Europe, was cleared first; secondly, a middle region, half wood, half arable, as at Moscow, etc.; thirdly, the present forest region, all in the north."

"The event which charmed me was the great Court ball of Wednesday, on the occasion of the marriage, to which we were invited by his Majesty's order. The entrances to the wonderful Winter Palace are so numerous that you are not surprised when you perceive how a thousand star-and-gartered eminences and well-dressed women have all within an hour found their way into the ' Salle Blanche.' The whole of this exquisite Palace being re-built and re-gilt, it is now in full beauty, and the blaze of light, the elegance of the candelabras, and the masses of gold, quite rivet attention. *We* have no notion of lighting, and I now understand the criticism of the foreigners who attended our Coronation.

"We waited for our presentation, which took place in about half an hour, when the Emperor came up to Lord Clanricarde, and asked for me, saying to me, ' You have travelled a great deal in our country, and intend to do so again.' On my thanking his Majesty for the kindness of my reception, he cut me short by saying, ' C'est à vous que nous devons nos remerciments profonds de venir parmi nous pour nous éclaircir et de nous être si utile. Je vous prie d'accepter mon personnel,' etc. He then asked if that was not my companion near me, and De Verneuil had his talk; but my excellent friend being short-sighted, had mistaken the Emperor, so that when his Majesty left us, De V. turned to me coolly and said, 'Eh bien ! c'est un homme très agréable que ce Grand Duc.'   'Mais c'est l'Empereur, mon cher ! '

"It was however in the advanced part of the evening that I really became intimate with the Czar.   I had glided through all the apartments, and was seated in converse with Count Strogonoff, when the Emperor appeared, and we were all on foot.   He selected me, and leaning against a pilaster began a regular conversation, asking me my opinion on various parts of the country.   After I had told him where I had been, he said, 'Great traveller as I am, you have already seen large tracts of my country which I have never visited.'   He then got me to open out upon my own hobby, and put me quite at home ; I ventured on my first endeavour at explanation, by stating how dearly I was interested in the structure of a country the whole northern region of which was made up of strata which I had spent so many years in classifying and arranging in other parts of Europe ; how their vast scale in Russia had surprised me, and how they offered evidences which were wanting in the western countries.   We then talked of coal, and I ventured on a geological lecture in order to explain where coal would not be found, the uses of our science, etc.   I ushered it in by saying that I was certain that his Majesty liked to know the truth, and my honest opinion, and he instantly said, 'Surtout, parlez franchement.'   Having given him the Silurian reasons against any coal deposits worthy of the name in any of the very ancient rocks on which his metropolis was situated, and a general view of the A B C, to all of which he listened most attentively, I then comforted him about the great coal-field of the Donetz, in Southern Russia, to which I was destined to go.   'Coal,' I said, 'was to be looked for in the south, and not in the north, which seemed a providential arrangement, as the forests were still plentiful in the latter, but annihilated in the former tracts. ' Ah!' said he, ' but how we have wasted our forests ! What disorder and irregularity has existed ! It is high time to put a stop to such practices, or God knows what would have been the state of the Empire, even under the reign of my son !' I then offered a few words in favour of the Crown peasants of the north, against whom the wood-cutting remark was directed, and spoke of their intelligence, honesty, and the absence of all great crimes, and how it had astonished us to travel through so wide a space, sleeping with our doors open, and in lofts or where we could, without being robbed, and in tracts where no soldiers or police existed. ' Oh !' added he, ' we are not however so savage as to allow such things.'

" After asking what was to be the length of our next tour, and what we hoped to find out and see, he desired me to express every wish to his officers, and all my wants should be supplied.

" He inquired about my former career, in what arms I had served, where and when, whether I was married, whether my wife ever came with me. On my saying that the day was when you were always at my side, and sketched and worked for me, he added, ' C'est ainsi avec ma femme, mais hélas sa santé ne le permet plus, elle a eu quinze couches.' Thus he chatted away, and talked of his children, and the happiness of his social circle.

" On my saying that I had served in infantry, cavalry, and staff in Portugal, Spain, and Sicily, his Majesty evidently took to me, for he said that his doctrine always had been that the army was the best school for every profession, and he was right glad to see that it made a good geologist. I then expressed how strong a desire I had to see the Russian army, adding that I had been out at six in the morning in the Champ de Mars, and had already seen his Majesty working some regiments of cavalry. 'What!' said he, 'talk of that morning drill; we were all dirty and not fit to be seen: to-morrow you shall see us better.' And then calling General Benhendorff, ' Donnez un bon cheval à M. Murchison pour la Grande Parade.' He then added, ' Mais c'est à Moscou que vous deviez nous voir parmi nos enfants—c'est ainsi que l'Impératrice et moi nous appelons nos Russes.'

" He talked with favour of his good English friends, and how well they had always served him. 'Alas!' said he, ' we have just lost two in the space of a few days, and on Friday we bury Admiral ——, an excellent officer and a very brave man, whom I greatly regret.'

" Two days had passed, and amidst my thousand occupations I had forgotten the Emperor's words. On Friday morning, when in my dressing-gown, *à la Russe*, at breakfast, the son of old Mrs. Wilson, our landlady, rushed in exclaiming, 'La, mother, only think of it! At eight o'clock the Emperor came in a single drosky to the English Church, and had to wait I know not how long before the parson came, and then he went through all the ceremony.' The old Admiral, being a Protestant, was buried in a vault under the English Church. I then bemoaned my want of tact in not having had my uniform on and ready at the church to meet the great man who thus honoured the memory of my countryman."

The letters and diaries written by our traveller at this season of rejoicing contain records of little else than the names of the great folks at whose houses he dined, or whom

he met at the Imperial entertainments.   During the day he
seems to have found time for an occasional interview with
some of the scientific men of St. Petersburg, and for desultory preparation for his journey.   But evidently courtiers
and court life had for the time quite dispossessed geologists
and geology in the attentions of the author of the " Silurian
System."   At the beginning of the week following that in
which he had made the acquaintance of the Czar and Imperial family, he attended a ball given by the newly married
Czarewitch.   From his reminiscences of that evening a few
sentences may be quoted.

" The Emperor talked to me again, asking me what I had
been doing this morning.   'Four hours,' said he, ' at the
School of Mines, and two hours with Professor Eichwald !
Why, you will quite tire yourself before you set out on your
long journey.   You must have good stout legs,' he continued,
passing his hand at the same time to the side of my thigh,
which he pinched.   He then discoursed of discipline, system,
etc., and alluding to the review of the morrow, he observed,
You will see three of my sons in the corps of the cadets.'
'The Grand Duke Constantine will, I suppose, command
them?' said I.   'Command!' replied he.   'No, indeed! he
will not even be in the front rank of privates ; he is yet
too young.   The little fellow has plenty of talent, but
requires to be kept in order.   We must have a good bridle
on him for some time to come.'   His Majesty again spoke to
me with gratitude concerning my labours, and said he had
no doubt my success in my present profession was mainly
due to my old military education, which he thought was the
best school for all men.

" The balls, parties, and reviews attendant on the Imperial

marriage being over, it was time to take to real work, and to
begin the geological researches on the grand scale which had
been devised through the departmental activity of General
Tcheffkine, then serving under the Minister of Finance,
Cancrine, and being chief of the School of Mines."

Count von Keyserling was named by the Imperial Government as one of the geologists of the expedition, with the invaluable Lieutenant Koksharof, who was again appointed to
accompany the travellers, and smooth their way for them.
The plan of operations embraced a series of traverses of the
vast central and southern provinces of the empire, together
with as full an examination as could be made of the chain
of the Ural Mountains. The party was to divide for short
periods, and meet again at given points, to compare and continue its observations, with the expectation of being able, perhaps, to concentrate the work of even two summers into one.

" All our inspections of collections, schools of mines,
academies, etc., being over, and our notebooks filled with
memoranda of things to be seen in Russia in Europe and the
Ural Mountains, there was still one grand public fête to be
witnessed. The Emperor, as Cancrine had reminded me,
had asked me to see him among his true Russians at Moscow. But this was not to take place for a fortnight, and in
that time the geological division under my orders might
effect much. So we galloped away to Moscow."

Their object was to examine the various outcrops of
limestone and thin seams of coal south of Moscow—a task
which was successfully accomplished without any note-
worthy incident. Up and away to their labours, sometimes
by three o'clock in the morning, the travellers contrived to
get over a goodly number of leagues of country, and, rattling

over the ground in their tilega, to raise many a thick cloud of dust from the "Tchornaia Zemlia" or black-earth of the Russian plains, so that they returned to Moscow in a sadly-begrimed condition, but in time for the fêtes.

"The great event of the Emperor presenting the heir-apparent to his people was about to come off. At 10 A.M. we drove to the Kremlin. We were ushered through crowds of Russian officers in the palace, and eventually found our way to the top of the building. I was on the balcony, close to the room whence the Emperor issued. He observed me, and nodded to me. At 11 he issued on foot and descended the steps in full Cossack dress to the Grande Place, which he had to cross to reach the great church, and at least 20,000 persons now filled it. A very narrow way had been formed up to this moment, but when the great bell tolled and Nicholas issued forth to the threshold, all line was broken, and the crowd presented itself in one dense mass before him like a wall. He stepped down towards them, and some touching his clothes, others his hands, he waved his hand gently up and down, and the dense mass opened out before him. Like a wedge he worked his way through the adoring multitude, who were clinging round his legs and touching his clothes. . . .

"Profiting by Demidoff's kindness, by half-past twelve we finally stormed the Kremlin, and forced on into the central tower, where we placed the niece of Napoleon [the Princess Mathilde] between De Verneuil and myself, like a Princess of the Kremlin, M. Demidoff acting as her Russian marito, and we as her French and English aides-de-camp. We were destined to wait for the great sight an hour or two, during which excellent sandwiches and good Madeira and sherry, and the French conversation, full of naïve and sparkling sallies from the daughter of Jerome, made us pass the time most agreeably. At length the cortége arrived—the good Marie in her calèche and four greys, the Emperor on her right hand, her brothers on the left, and the Grand Duke Héritier passing close along the line of troops. When they entered the Holy Gate of the Kremlin, the sight of course closed for us.

" As we descended the staircase, thinking all sights were over, the attendants stopped us at a doorway, and, in an instant, the Emperor, with the Grand Duchess on his arm, passed within a few paces of us. He at once recognised us with a gracious nod. Of this I should not have felt so certain if Count Benhendorff had not told me two hours afterwards that his Majesty had informed him of our position. Nicholas's eye is everywhere, and long may it be so !

" Count Benhendorff gave us an account of the Imperial reception. At Ribinsk—a thriving commercial town on the Volga, with 30,000 inhabitants—it appears that the people who had never seen the Emperor kept up such a roar under the Imperial residence, that at last, when midnight came, they were requested to allow the Emperor to sleep. The hint was no sooner given than obeyed. But what followed ? Not a man slunk sulkily away ; the loyal mass lay down and slept at their posts till the return of day was ushered in by a general chanticleer from those sturdy monarch-loving Muscovites. Well then may Nicholas exclaim, 'These good people are not yet so advanced as to have learnt not to love their sovereign '—words which he used to me in speaking of the Russians of the interior.

" Benhendorff also informed me that the horse-artillery

which we saw this morning had marched 110 versts the
day before, *i.e.* seventy miles ! This beats the famous march
of the old Fourth Dragoons, my father-in-law, General
Hugonin's regiment, which marched from Canterbury to
London in a day, and acted that evening in the Borough in
quelling one of the Lord George Gordon riots in 1784."

But it was now time to doff uniforms and court-dresses,
and take to the more homely garb of travelling geologists.
Murchison and his friends had planned their journey in such
a way that it should comprise many minor lateral excursions, and they now proceeded to put the plan into execution. Starting from Moscow, they crossed the empire by
Vladimir, Kasan, and Perm into the Ural Mountains, and
the edge of the vast steppes of Siberia. From these remote
bounds they turned southwards to explore the southern
Urals as far as Orsk, whence, bending their course once more
in a westerly direction, they passed through Orenburg, re-
crossing the Volga at Sarepta, traversing the country of the
Don Cossacks to the Sea of Azov, and then turning northward to make another traverse of the empire back by
Moscow to St. Petersburg.

Five busy months passed away in these journeys. Murchison kept as usual a full diary. Being mainly geological,
his memoranda were subsequently elaborated into the great
work on " Russia and the Ural Mountains." But among
them occur records of incidents of travel and other notes,
which give us glimpses of the scenery and people among
whom he lived, and of the way in which this extensive and
rapid survey of the Russian domains was achieved.

As on the previous journey, the main highways of the

country were followed. Provided with a formidable Imperial document, countersigned and double-sealed to enforce attention from all persons in authority along their route, the travellers had usually little difficulty in procuring horses at the stations.   In most cases, indeed, the chief dignitary of each place waited on them personally, and in not a few instances treated them with the frankest hospitality.   The kindness which Murchison experienced in this way even in the wildest tracts of the empire, filled him with that deep affection for Russia and the Russians which used to show itself continually all through his life.   But neither Imperial ukase nor kindly proffered assistance could wholly overcome the natural difficulties of the country.   The geologists had made up their mind to a good deal of rough fare and sorry lodging, nor in these respects were their prognostications unrealized.

During the earlier part of the journey through Vladimir, Nijnii Novgorod, and Kazan, there was little either in the geology or the scenery to delay the expedition.   Murchison, indeed, seems to have got so disgusted with the interminable red sandstones and marls as to break out into some doggerel lines in French, that being the language which was now his only mode of communication with his travelling companions and the natives of the country.   These rocks were not yet understood by him.   He became proud enough of them before long, for they furnished to him the type of a new geological subdivision, to which, from the province where they were so well developed, he gave the name of " Permian."

In spite of these tedious red rocks, Kazan afforded some interest.   The fat jolly Vice-Governor had instructions to look well after the travellers, and it would appear that he did his

best. In their honour he donned his full uniform, white
laced hat, and numerous orders, and arrived at their inn with
the determination that they should see everything in Kazan
forthwith. In vain they explained that one of the Professors
had already kindly offered to escort them through the collections of the University. What! had he not received the
Imperial command to look after them himself? and besides,
had he not been a sailor in the days of the old war, when
the British and Russian fleets were allied, and did he not
still remember a few broken words of English—" I beg you,
sare," "ver much wind," etc.! He would show them the
collections, and everything and everybody too. De Verneuil
and von Keyserling had made a detour. Murchison, therefore,
under the supervision of the Vice-Governor, took further
notes for the Ural Survey from the specimens and information obtainable from the Professors, and attended sundry
feasts into which the exuberant hospitality of Kazan broke
out. When the party reunited, and all was ready for the
march again, the Vice-Governor must needs give one farewell banquet. " We sat down," Murchison writes, " forty-
five in a small room, and the Vice-Governor was quite
charming with his old sailor-loves of ' Sally Cox ' and
' Mary Dickenson ' when in England."

Over many leagues of red rocks the party journeyed
through the government of Perm towards the long low
ridges of the Urals. They passed on the way a gang of
manacled prisoners bound for Siberia, to whom, amid his
notes of " Roth-todt-liegende," " Nagelflue," and other geological matters, Murchison devotes a few words in his journal.
About a hundred and fifty men and women, under a strong
military escort, the men in some cases manacled in couples,

were marching to their exile. "Thank God!" he writes, "in England we have the sea for our high-road to banishment; for such scenes are very harassing."

While the exiles were tramping along the highway, the geologists, having gained a rising ground, were luxuriating in the first distinct view of the real crest, if it may be so called, of the chain of the Ural Mountains—a long, slightly undulated line, rising behind a succession of wooded ridges, and forming a singularly unimpressive landscape, considered as a part of one of the leading mountain-chains of Europe. It was not easy to say when the mountain land was really entered, so gradual had been the ascent. "Though the Ural had been a chain in my imagination, we were really going over it at a gallop, the highest hill, indeed, not exceeding (in elevation above its base) our Surrey Lower Green-sandstone." With no rocks on either side of the dull road, and with dark rainy weather, the passage of one of the depressions in the low watershed of Europe and Asia became dreary and monotonous, till the travellers found themselves in the heart of the gold-mining region and in a comfortable inn at Ekaterinburg.

Over vast tracts of Russia the rocks lie in horizontal sheets, so little disturbed that, failing river gorges and other natural sections, it becomes no easy task to determine their proper order. Like a series of sheets of cloth laid on a table, the uppermost conceals those which lie beneath it. Eastwards, however, they have been ridged up into the long swell of the Urals, and our travellers, having already acquired a good deal of miscellaneous information from the labours of Humboldt, G. Rose, Ehrenberg, Helmersen, Hoffman, and others, regarding that little-known tract, were now bent upon discovering how far the elevation of the Ural chain had exposed the edges of the strata, so as to allow their order and thickness and fossil contents to be determined in an easier and more satisfactory way than could be done over leagues of the flat lowlands.   They lost no time in beginning their work, and before many weeks had passed, by dividing their forces into two parties, and moving upon separate but parallel lines of research, with occasional reunions by converging traverses at the chief mining establishments, they succeeded in ascertaining the general geological structure of the Ural Mountains, in such a way as to permit the main masses of the rocks in that chain to be effectually compared with the geological succession already established elsewhere in Europe.

One great impediment in their way was the want of any even tolerable map on which to record their work—a want, the paralysing effect of which only the geologist who has been similarly placed can adequately appreciate.   "Were I Emperor of Russia," he writes, "I would make verily at least one thousand of my lazy officers work for their laced coats, and produce me a good map, or they should study physical geography in Eastern Siberia.   Excepting General Tcheffkine and a few, very few, I never met with any man who knew how to handle a map.   It is really an affair of an hour to get a governor to make his way upon a map along a well-beaten road.   I never shall forget my surprise last year at Nijnii Novgorod, when the Government House was ransacked for a map, upon which my line of march to the south of Moscow was to be traced.   At length what came forth from this centre of Russian wealth and commerce, in the very fair of Nijnii, and in the Government House?—A

district map of Schoubert's which I have so anathematized?
—No, but one of the little three-rouble maps which the
common traveller buys, with simply the names of the chief
places and small towns! The same occurred at Kostroma,
where the Governor had no other.

" If such be the case in the heart of Russia, how are we to
expect that the best-informed natives here in the Urals
should have any idea of their broken and diversified region?
Russia must produce geographers before she can expect to
have geologists. The cost of a single regiment of cavalry
would effect this great national work; and would that the
Emperor could be led to see its desirableness and efficacy for
all good measures of internal improvement! I never yet
heard a Russian speak of any place as being east, west, north,
or south of such a point, but merely as so many versts from
this or that town. Ask him in what direction and he is
dumb. First he will say it is to the right or to the left,
according as he may have travelled; and it is only by a
serious cross-examination, which would puzzle a barrister of
the northern circuit, that you can guess at something like
the fact. But alas! after fancying myself informed, how
wide have I found nature from their mark! Here, for
example, you will find people disputing as to whether a
leading place, such as Stataoust, is to the east or west
side of the Ural; and as for the roads, they trust to
their clever peasants, stout horses, and ever-resisting taran-
tasse."

The absence of reliable maps, though it proved a continual hindrance in the process of geologizing, was never
allowed to retard the bodily activity of the party. Of that
party and its local auxiliaries, as they started on one of

their exploratory tours, the journal gives the following
account :—

"A route from the Zavod [mining-station] of Chresto-
vodsvisgensk across the Ural chain to the valley of the Is,
on the eastern watershed, was now to be undertaken, as
arranged in our programme. But this was no slight
affair, inasmuch as no party had travelled by this old and
abandoned corduroy road through the forests and sloughs
for many years, yet, by sending peasants across, arrangements were made.

"At 3 A.M., 2d July, I roused the whole party, and at $\frac{1}{4}$
past four we were in march from the Zavod, being a party
of twenty cavaliers of most grotesque and varied outline.
The President of the Geological Society need not describe
himself. The Vice-President of the Geological Society of
France sported his long blue Spanish cloak, and a broad-
brimmed, round-topped, Moscow grey hat, which, on the
back of a Wouvermanns' grey horse, formed an essential
item in the motley group. Herr Graube, the Master of
the Mint, who led us, had his long boots above his knees,
and large furred coloshes, with his little German cap. Von
Keyserling, in his green cap and jacket, bestrode a gallant
brown, and his servant, Juan the Venerable, turned out on
a Russian saddle in a long black cloak, on a white Cossack-
like beast. The Ispravnick of the district, who honoured
us, was a sort of sub-military looking figure, with spectacles
and Life-guard boots, superadded to a black shooting-
jacket. The German doctor of the Zavod, a most obliging
man, was mounted on a capital iron-grey, with high action.
Lastly came our two Russian officers, Karspinski and
Koksharoff, both of whom were knocked up by our rapid

ride of yesterday. The former, dreading the result, to-day had strapped a large pillow on his russia-leather red and yellow demi-peak saddle. Our bearded fellows were perhaps the best for the painter, with their caftans, double-coned hats, and long boots; one armed with an axe behind; another with De Verneuil's gun in hand; a third with long Turkish pipes; and others astride of animals carrying sacks, bags, and beds.

"Our start was somewhat cheerless as to weather, for the day looked lowering; and in a few minutes we were in the interminable boggy forests which fringe the flanks of the Ural. It was soon evident that all haste was in vain. The sloughs exceeded all that my imagination had conjured up. The road was a sort of bridle-road, not to be described to English understanding, for it consisted in most parts, and for ten or twenty versts, of planks and round trees, most of them rotten and breaking, placed over the quagmires here and there, the track along which seemed hopeless, but for the dexterity of a Russian horse. If the plank broke and his leg went in up to the hock, he pulled it leisurely out, whilst with the other he was fighting his way up the rounded slippery single plank which remained. If his tread on one end brought the other up in his face, he would gently and evenly move on till the equilibrium was established, and he gained another safe footing. Add to this, massive trees, including the noble *Pinus cembra* and others, lying across the road, immense roots branching in all directions, sedge and long grass up to the horse's belly, and you may have some idea of a bridle-road in the Ural."

Not much geology could be done under such unfavourable

conditions, nor could any clear notion be formed of the
general aspect of the Ural chain, though the peculiarities
of the wooded regions became only too familiar. Now and
then the travellers succeeded in getting above the line of
wood, so as to catch a glimpse of the summits of the
Ural and the country beyond. Thus at the Katchkanar
they " at last found a true mountain in the Ural "—rough

View from the Summit of the Katchkanar, North Ural, looking northwards.
(From *Russia in Europe*, vol. i. p. 392.)

splintered crags, shooting high over the damp sombre forests,
and nourishing in their crevices and amid their slopes a
bright and luxuriant vegetation which recalled that of some
Swiss valley. From this peak they could look on one side
over the far rolling sea of dark pine, with here and there a
snow-streaked summit rising island-like out of it ; on the
other side lay the vast plains of Siberia, with the level

featureless surface, and to the eye at least with the boundless horizon of a great sea.[1]

At other places on the crest of the chain rocky scarps
were encountered. From Stataoust the party reached some
conspicuous rocks rising along the water parting between
Europe and Asia. "Clambering up to the summit, and
with one leg on either Continent, we sang 'God save the
Emperor.' In this sequestered spot, however, neither officers
nor workmen knew the present national air, which I had
heard at St. Petersburg and Moscow, but began to chant
our old 'God save the King,' which they had sung since
the time of Peter the Great. I then hummed this new
air, and this music of Levoff was thus first given out in
the western borders of Siberia."

But the most exciting and instructive work which they
carried out in these remote regions was the exploration of
some of the river-courses. Owing to the need of abundant
water-power for mining purposes, the streams had been
manipulated in many different ways, some being turned
into a succession of dams and waterfalls, others deprived of
their water to fill lateral reservoirs. It was in these natural
sections that the true structure of the Ural might be most
confidently searched for, and special care was given to them,
though but for the active co-operation of the mining
authorities, these defiles would have proved far more formidable obstacles than the morasses and corduroy bridle-
tracks. How the work was done may be judged from the
following extract :—

"Descending the river Issetz in canoes, between rocky
banks of micaceous schists and granite, we came to the

[1] See Plate, p. 392 of *Russia and the Ural Mountains.*

mill of Paulken, where the miller offered us tea, observing
that his first love was God and the Emperor, the next
strangers; for he had travelled in Russia, and knew the
value of hospitality. The descent of this river is quite
unique, for the water-traveller must quit his canoe at every
one of the hundred mill-races. There are upwards of two
hundred of these mill-dams between Ekaterinburg and
Kamensk. At every one of these, one's goods, chattels, and
self must go out and in, and his canoe be shoved over the
rough roots, sticks, and blocks (often held together by
large blocks of stones), and dropped some eight or
fifteen feet as the case may be. No ordinary traveller
can execute this journey without great loss of time and
patience. For us the authorities were so active that at
each stoppage a multitude was waiting to get us through.
The sub-officer put every ' starosta ' in play, and our descent
was a regular press. ' Stupai, pikarea, poshol ! ' and on we
went (at what cost it matters not in this land), carrying
with us the inmates of one village till we reached the next.
No one who has not descended this Siberian river would
believe how much comfort and industry appear on its banks.
No mill, numerous as they were, was without six or more
little carts before it. A dense population lives all along
the Issetz. Good white large churches rise up here and
there, and everywhere the cottages are nice and clean."

More adventurous was the descent of one of the streams on
the other or western slope of the Ural. Von Keyserling and
De Verneuil had been making independent observations, and
the party re-united at a mining station on the Serebrianska,
a small stream flowing into the Tchussovaya, which descends
into the great Permian lowlands. " The descent of the Sere-

brianska," he says, " was one of the most memorable days
of my life. The distance to be accomplished by this winding
stream was seventy versts, or nearly fifty English miles.
When I went to rest, the bed of the river was almost quite
dry, with not water enough to drown a rat, and yet we were
to effect the miracle of floating down in a six-oared boat.
When I awoke a furious stream was rushing down, and the

Lake of Aushkul, South Ural.—(From *Russia in Europe*, vol. i. p. 359.)

natives were beginning to get canoes. The good comman-
dant, having the Imperial order that I was to descend by
water, had let off an upper lake, and thus made a river in
a fine dry sunny day !

"The waters having been let off for us, and the river bed
filled, we effected our embarkation amid three cheers. The
river was muddy, and had rocks hidden, with very sharp

curves of the stream. With a hundred groundings and
stoppages, we got tired of our big boat of honour, and took
to the canoes. These answered well for a while, but trusting to shoot through some stakes and nets (myself on my
back at the head of the canoe), we (*i.e.* De Verneuil and
myself) were capsized in a strong current. I saved my
note-book (see the stains), but my cloak, bag, pipe, etc., went
floating down. A curious scene followed, after we had
scrambled out to the shore. The other canoe shot by and
picked up our floating apparatus. Fortunately this letting
off the waters had brought down some natives to catch fish,
and they had a fire, by which we dried ourselves, whilst
their large wolf-dogs lay around us. When we re-em-
barked, we shot several ducks (Merjanier), and here and
there found limestones and shales striking to the N.N.W.
Some of the limestones were charged with Devonian fossils.

"After this, evening began to fall. Saddles, anticlinals,
and synclinals arose in magnificent masses on the rocky
banks, but our boat-bottom was soon knocked to pieces by
grounding at least a hundred times, and whisking round as in
a waltz at each shock. It now filled so rapidly that we had
just time to escape. We had then a fine evening scene.
We landed on shingle, and got into the forest, not having
seen a house or hut for fifty miles. The dense wildness of
the scene, the jungle and intricacy of a Russian forest,
can never be forgotten. We had to cross fallen trees and
branches, and to force through underwood up to our necks.

"After our various night evolutions, sometimes by land
and sometimes by water, we finally reached our ' derevna '
(Ust Serebrianska) at two A.M., wet up to the middle, by
walking through moist jungle and meadow. Our men were

very amphibia, and required no food. They had been half
the day in that stream, pulling, hauling, shoving, and shout-
ing, and never eating or drinking. We had to awake the
chief peasant's family, and were soon in a fine hot room,
with children sleeping all about.

"I awoke with the bright sun, after three hours' rest, and

Gorge of the Tchussovaya, west flank of Ural. Contorted Devonian and Carboniferous
Rocks.—(From *Russia in Europe,* vol. i. p. 386.)

pulling my shoes out of the oven, and my dried clothes
from the various long poles, proceeded after a warm tea to
embark on the Tschussovaya, into which the Serebrianska
flows. The Tschussovaya being a much larger river, we had

no difficulty in boating down it, and we had a most instructive and exciting day, as we passed in the deep gorges of
Devonian and Carboniferous limestone, here thrown up in
vertical beds to form peaks, then coiled over even like ropes
in a storm, or broken in every direction. Making many
sections, with many memoranda, the 17th June was
finished."

" On the following day we worked away down the river
in the same great leaky boat as before, the boatmen singing
their carols, and abusing the Ispravnicks and proprietors
who force them to drink bad ' vodki' or whisky by their
monopoly.   Other songs were gentle, plaintive love-ditties,
so unlike what our coarse country fellows would sing.
With no stimulants, getting but black bread, and working
in wet clothes, for they were continually in the river shoving
the boat on, they sang in rhymes, one of which as translated by Koksharoff was :—

  ' My love she lives on the banks of a rapid stream,
    And when she goes to the garden to pull a rose, she thinks of me.'

Another of these ditties began—' Mary, come back from
the bower.'   A third was a comic song, quizzing a soldier
who got into a house when tipsy.   A fourth was a jollification of peasants in a drinking-shop, to beat the maker of
bad brandy, with a famous loud refrain in which all the
boatmen joined heartily."

When, after toils of this kind, the travellers found themselves again in one or other of the busy mining stations, they
met with much courteous, and even exuberant, hospitality.
Thus before leaving Ekaterinburg a dinner was given in their
honour, to which the chief officials of the place were asked.
Delicacies of all kinds, as well as costly wines, appeared at

the table. "The dinner," says Murchison, "finished by a bumper of champagne to my wife, and throwing all the glasses out of the building, that they might never again be used. I made a speech in reply, and begged to have a top and a bottom of the broken glasses, that I might reunite them with a silver plate in England, and inscribe on it my grateful thanks."

Posts were neither frequent nor regular, or at least the geologists were too constantly on the move to be able to count upon many fixed addresses to which letters could be sent for them. Murchison, however, though busy, body and soul, in Russian geology, naturally found his thoughts many a time far away among his friends at home. On 28th July, by four

Plain of Limestone in the South Ural.—(From *Russia in Europe*, vol. i. p. 439.)

in the morning, he was up, had boiled his own kettle and breakfasted, and was writing up his journal notes :—"This day the British Association is assembling at Plymouth, and I drank success to it. How few of the members there will have lighter hearts than their general secretary in Siberia ! . . . . In this poor dreary spot (for the Steppes are like the flat border counties of England and Scotland) I made two children at all events right happy by giving them new large copper pieces."

It was in the southern parts of the Ural that the travellers had most experience of those grassy plains, to which the term Steppes is applied—"wide, monotonous,

featureless plateaux, the withered grassy surface undulating
to the south and west, while to the east all is boundless
even.   Not a glimpse of what may be called the Ural
mountains.   The country becomes more decidedly southern;
or, in other words, bare, barren, and bad.   Dried dung, piled
up, is now used in place of wood, and Kirghis and Calmuck
faces appear under the military uniform in very poor villages.
The road now quits the low eminences on which the station
is placed, defended by men of all arms, including Cossacks,
and passes along the wide sea of the Steppe.   Low bushes
of a sort of *Myrica* are mixed with a little culture of oats and
corn.   The very road was grassy, and we galloped by the
first armed mounted archer Bashkirs I had seen, with a
stout double bow, and twenty heavy arrows.   They are used
in protecting the conveyance of goods."

Notices of some of the most striking features of the tribes
through which the journey led occur in the journal.   "Our
Bashkir drivers had a name for every hill, however small.
The principal man, or coachman, was a fine long, aquiline-
nosed, wild-looking, good-humoured fellow, with a cap of
loose shaggy fur.   He had the three wheelers in hand, preceded by two postilions with a pair each, and all these were
headed by a long lad riding a leader in advance.   Our
equipage and ponies measured fifty feet in length.   The
Bashkirs, being accustomed only to horseback, are not good
whips like the Rushki, and their horses are too weak to
charge a hill; but they go down one furiously,—no slight
danger for the riders, and for us also, who, in case of a fall,
would have been well smashed."

These Bashkir of the Ural had no sympathy with the
geologists in their search after the mammoth and other bones

found in the gold-drifts and ancient alluvia of those regions. " These they considered as relics of their great forefathers, saying, ' Take our gold if you will, but leave us, for God's sake, the bones of our ancestors !' "

One hot day the party arrived at a little station in the South Ural. "Dined at this lonely spot. All still as death at noon. Grasses all burnt up. People asleep, but soon awakened. The Cossack women of the Uralsk are fine broad creatures in red dresses. The confidence of these primitive people is very great, for they allowed us to grope for tea-spoons and bread in the cupboards in which their bank-notes and roubles were lying loose !"

Living in Bashkir tents, the geologists learned to relish a sort of diet which anywhere else might have been deemed hardly tolerable. One staple article of food in summer among these simple people is " Koumiss,"—a preparation of mare's milk,—" the staff of life, the bread, meat, and wine of the Bashkir." Of this liquor Murchison would appear to have become fond, and to have thriven on it. He tells how at one of the Bashkir stations, where the party had spent the night, " after a very good breakfast, all sorts of saluta-tions followed, such as the drinking of Koumiss to the prosperity of our host. Then we heard his story of losing sixty sheep, killed by three wolves last winter; next we found that he paid so many roubles for his present wife, and that her dress cost him more than herself. I expressed a wish to him to have a Bashkir vest, belt, pouch, and cap, and he offered me his own. It was with difficulty that I got him to take the value to replace them."[1]

[1] " This dress I afterwards wore at a fancy ball at Stafford House, when I saluted the old Duke of Wellington in true Bashkir style. Not

At last, with note-books laden with descriptions and sections of the various traverses which they had made of the Ural chain, the travellers began to move once more into the great western plain.   They had succeeded in reaching the central masses of that chain, and in recognising, by fossil evidence, that from a nucleus of granite and crystalline rocks, Silurian, Devonian, and Carboniferous strata are successively thrown off.   This evidence had been industriously gathered from river-channels, road-sides, mining operations, and every available source of information.   For days together they had been off soon after daybreak for renewed hammering, and many a time night descended upon them while they were still plying their task.   Now and then, indeed, when pinched for time, they even essayed to use their hammers in the dark, after the manner of M. Boubée, whose example Murchison used jocularly to quote, up to the end of his life.[1]

It was now time to turn westwards, towards the coalfields of the south of Russia, the exploration of which had been fixed as one of the chief objects of the expedition. But Orenburg lay in their way, with its governor, the brave, though unfortunate, hero of the Khivan expedition, General Perovski.   He was then at his country quarters, in a picturesque wooded valley at the far edge of the Steppe, a long way to the north-east of the town.   To see a little more geology, with a taste of Russian sport, and the

one of my intimate friends recognised me.   The sword, etc., I had from Stataoust, and medals *à la Russe*, hung round me."

[1] This geologist, said Sir Roderick, used to maintain that a good deal of geological work could be done as well by night as by day.   Rocks had three well-marked sounds under the hammer—*Piff*, *Paff*, and *Puff!* The first of these indicated the hard crystalline rocks, the second the sandstones, and the third the clays !

acquaintance of a noted Russian soldier, were attractions Murchison could not resist. So he undertook the interminably tedious drive across the Steppe, and spent a few days with more thorough pleasure than he had enjoyed since leaving home. With all the comforts of civilized life, this place was yet quite in the wilds,—Bashkir attendants, with their picturesque costumes, a blazing bonfire lighted in the encampment, and the moonlight glancing on the lances of the Bashkir guard. Perovski made a great impression on the retired officer of the 36th. One evening he gave him the following anecdote:—"When the utter failure of the Khivan expedition become known, all Russia turned upon me, and with any other master than my good Emperor I was a ruined man. But the Emperor declared he would not condemn me until the opinion of the Duke of Wellington was obtained, who, being a Marshal in the Russian army, should have the whole case laid before him. This was done through Baron Brunnow, and then came the Duke's dictum : ' I am of opinion that General Perovski acted as a skilful general, and that if he had not retreated when he did, instead of losing a fourth part of his army, he might have lost the whole. Success was impossible under such intense cold.'" On this judgment being given, the Emperor not only absolved Perovski, but gave him the government of Orenburg. The General added,—"You see that I owe everything to your illustrious Duke, and I beg of you, when you return to England, to take some opportunity of letting him know what a grateful person I am." "This," Murchison adds, " I took care to do."

The visit to the General led actually to yet another traverse of the Ural, for he showed the travellers a map of

the southern part of the chain, so greatly superior to anything which they had yet been fortunate enough to meet with,
that it prompted a strong desire to take one final look at the
Ural geology, and with his help among the Bashkir population, they succeeded in once more crossing the chain in its
central part, and collated their work in the southern and
northern portions.

At last, however, they had unwillingly to turn their

The Gurmaya Hills, South Ural, approaching from the Steppes.
(From *Russia in Europe*, vol. i. p. 450.)

backs finally upon those picturesque ridges and fertile
valleys of the Southern Ural, and to speed westwards
through the dreary monotonous country of the Steppes. In
geology there was nothing either very interesting or complicated to detain them. They therefore hurried on through
the Kirghis Steppes to Sarepta, crossing once more the great
Volga, and tracing as they went some of the limits of the

ancient sea of which the present Caspian is but a shrunk
remnant.   Through the plains of the Don, among Cossacks
and Kalmucks, their course was yet more rapid.   On 8th
September the journal records,—"De Verneuil sleeping in
the hut, and myself in the carriage.   What is a Cossack
post station?   Everything about it is very different from a
flaming great wooden Russian station.   First, you see a dot
upon the Steppe, which magnifies as you approach it to a
thing about the size of the smallest Irish hut, and not very
unlike one in externals, being concocted of mud and reeds,
with very little wood.   But the interior is very different
from an Irish cabin.   I now write in a room ten feet square,
and on the table lieth the regular sealed post-book.   This
official chamber is six and a half feet high, and has a large
stove in the corner, a door four feet high, and two windows
eighteen inches by nine.   The walls are all well white-
washed, the tables well scoured, and the floor well beaten
and clean swept."

Skirting the sea of Azov, they turned northwards into the
coal-field of the Donetz.   There they made a series of most
important observations, bearing both on general questions of
geology and on the industrial resources of the Russian
Empire.   They found the coal-seams to lie, like many of
those in the north of England and in Scotland, among the
marine strata of the Carboniferous Limestone, there being,
so far as they could see, no true "Coal-measures," in the
geologist's sense of that term, in Russia.   They learnt, moreover, that though the coal was quite workable, and had
indeed been mined for years, it lay among strata which,
unlike those of the vast tracts in the centre of the Empire,
had been subject to such underground disturbances as to

present many large dislocations and many foldings. They traced it westwards until they found it die out again on ancient crystalline rocks, while northward and eastwards they learnt that it passed under sheets of Cretaceous and Tertiary deposits.

In the course of this prolonged tour, while the main attention of the geologists had been given to the structure of the solid rocks, their ingenuity had been on many occasions called forth by the anomalous features presented by the surface deposits of the country. These difficulties started up in renewed force on the way north to Moscow. They are thus stated in the journal :—" The surface of Russia affords some puzzling problems. In passing from south to north you first meet with the tract of the northern drift, the materials of which become more and more numerous at every ten versts. Still the old rule (applied by me last year) answers perfectly, viz., the diluvia are three-fourths derived from the subjacent rocks, so as largely and loosely to indicate the zone of country you are traversing, provided you have the key to the subsoils of Russia. Thus, whilst the loose stuff was all yellow in the country composed of yellow Devonians, so to-day, viz., from Lichvin to Kaluga, you are immersed either in ferruginous, or reddish, or white sands. The latter prevail in great quantity in the horrible tracts north and south of Peremyschl—a most wretched town,—and their presence is well explained by the destruction of the yellow and white sands of the Carboniferous Limestone; for, with the exception of the section opposite to Peremyschl, and one or two rare localities, the valley of the Oka is here denuded to a width of several versts, which space is flooded in spring-time. This is one of the numerous cases which realize in modern times (viz., in spring-floods) the geologist's idea (mine at least) of the condition of the earth's surface during the intermediate period, viz., shortly after emersion from the sea, when the mammoth had left his bones sticking in the mud.

" The drifting and excavation are explicable as in other places. The vast spaces denuded and broken up in the most horizontal districts explain perfectly the vast masses of local detritus in the northern governments, and their transport for 150 versts southwards.

" But how explain the Tchornaia-zem which overlaps the diluvium of the north, and is also spread over vast regions of the centre and south of Russia, sometimes in river valleys, sometimes on slopes, sometimes on high plateaux, and is always of precisely the same composition, without a trace of true pebbles, or, in short, of any extra ingredient ? What colours the black loam ? If it be of vegetable origin, whole forests of mighty extent must have been destroyed to pro-duce it. But how destroyed ? In all other superficial deposits, whether in bog, in mud, or in the youngest tertiaries, we find traces of the trees, branches, grasses, etc., but not a vestige have we in the Tchornoi-zem. All is a black, uniform, finely levigated paste, sometimes highly tenacious, and very much so when not worked into with the plough, for after labour it works into a fine black mould. In this virgin state it is seldom to be seen, for 90 to 100 parts of all that is good in soil, from the Ural to the swamps of Poland, is already in culture. The specimens I selected, however, had evidently never been touched by plough or man ; they were taken from the precipitous sides of the Oka, just after a subsidence of the cliffs which exposed the section, the lowest deposit of

which is the iron sand which covers such large tracts in
Vladimir, and many governments, and overlaps the truncated
and denuded edges of the Devonian limestone in these parts.
Perhaps it is Tertiary, but only perhaps, for we have similar
ironstones under the chalk at Kursk, and similar limestones
over the Lower Jura shales at Saratoff.

" If the drift was, as I believe it to be, a great submarine
operation, then are we to suppose that the Tchornaia-zem is
the result of a great change of a pre-existing terrestrial surface ?   To believe in this seems to me very difficult, and for
this reason, that no imaginable destructive sub-aërial agency
could produce a general wide-spread and uniform condition.
By what conceivable sub-aërial agency can this very thick
black cerate have been spread out as with a mighty trowel,
and fashioned to the surface over millions of square miles ?
If forests were destroyed to furnish it, how were they so
triturated and reduced to this black cement, that no chemist
could invent apparatus to produce such results, even in a
crucible ?

" I end, therefore, in believing that this black earth is
the last covering of mud and slime which was left by the
retirement of the Liassic sea, and was to a great extent
derived from the wearing away of the shales of the Jurassic
strata [sic].

" If such are some of the difficulties of the Tchornaia-zem,
what are we to say of the great subjacent masses of clay
and sand of South Russia ?   In this we have not a pebble
of transport, nothing but a sort of clay or loam, which
might well pass for ' loess.'   If so, and if ' loess ' was produced as Lyell thinks, then all South and Central Russia
was one vast pond, in which all was tranquil during two

epochs—1*st*, that of the so-called drift, with mammoths;
2*d*, that of the black earth."

By the beginning of October the various members of the
party, who had separated for the purpose of making different
traverses of the country, were once more brought together in
Moscow.   There several days were spent by Murchison "in
condensing thoughts, comparing notes, examining Von Keyserling and Koksharoff, consulting with De Verneuil and all
the party, and preparing two general sections, a Tableau
Générale, the map, and the report of fourteen pages to Count
Cancrine on the results of the 'Expédition Géologique.' Also
a letter was concocted to old Professor Fischer, for publication
in the *Bulletin de Moscou* and the German periodicals, giving
a slight sketch of our doings, and in which I first suggested
the term Permian."   Petersburg was reached again on the
8th of October.

Of his last few days in Russia the journal records the
following memoranda :—" Having travelled 20,000 versts in
the distant provinces without losing a pin, we were twice
robbed between Novgorod and Moscow of our beds and
things behind the carriage.   One trunk only was left in the
hinder parts, and this was viced on; but besides this
security, I resolved to guard it from the station where we
detected our losses, and so letting down the head of the
calèche, I laid De Verneuil's double-barrelled gun over the
rear, and determined to bag the first thief who approached;
and in this form we reached Madam Wilson's house.
Besides several interviews with the old minister, Count
Cancrine (who was much gratified with my report, of which
he had prepared a digest for the Emperor), and a dinner at
his house, and the same at Tcheffkine's, we were occupied

in looking after more than twenty cases of fossils, which had arrived from our distant parts, and were deposited in the magazine of the School of Mines.

"All our reports and work being delivered in, official letters were received announcing the Second Class St. Anne in diamonds for myself, and a plain cross for De Verneuil, as a mark of the Emperor's approbation of our labours.

"We were to sail in the Nikolai steamer on Saturday the 24th, and Friday was fixed by the Emperor for seeing us— a great compliment, as it was His Majesty's working day with his ministers.   On these occasions Nicholas uses no ceremony.   After thanking us for taking so much pains about the Ural Mountains, and after asking if I thought the gold alluvia were likely to last much longer, he desired me to open out and explain the rolls of drawing and paper under my arm.   This I did *secundum artem*.   He was serious when he was receiving his lesson about the productive and non-productive tracts of coal, and the rationale thereof, and laughing when he saw the *Productus Cancrini* and the *Goniatites Tcheffkini* inscribed upon the roll, he asked, ' Quel espèce de produit est celui-là de mon ami le Comte ?' 'And so you have seen General Perovski?   He is my good and dear friend.   I hope you were pleased with him?' I had then to sing the praises, which I naturally did *con amore*, of the frank and gallant soldier who had been so truly kind, and also so very useful to us.

"When our geological talk was over, and he had asked us about our health, our travels, and many special points, I broached my desire to revisit Russia in 1843, with my work in my hand, and on that occasion to explore the Altai.

' Come when you will,' was his reply, ' I shall always rejoice to see you, and to afford you a hearty welcome; and be assured that I am most particularly grateful for all your exertions to impart knowledge amongst us whilst you are studying the natural history of our country.' And then with as hearty an ' au revoir,' and as warm a shaking of hands as ever took place between the oldest familiar friends, we took our leave.

" Such is Nicholas. Let those who criticise him look into his noble and frank countenance, and then let them try to tell me he is a tyrant. No ; utter ignorance of the nature of the man has led to this most unjust notion. Nicholas is above all deceit, and squares his conduct on more noble principles than that of any potentate of modern times. He disdains subterfuge, and is transparent as to all his emotions. Hence if ill-served (knowing perfectly what duty is) he does not suppress his feelings. He is sometimes quick in his anger, but like all such generous souls, his confidence in his friends is unbounded. Firm and unchanging in his resolves as an Emperor of Russia must be, if he desires to reign, his untiring aim is to ameliorate every institution which he can touch. But alas ! so bound up is everything in Russia by forms, customs, and prejudices, that he who supposes the autocrat powerful for all good, and capable of making every conceivable reform, would find himself most egregiously mistaken. The nobles and their privileges meet him here, the different bureaucracies there. Here the Minister of the State Demesnes places a veto upon some great projected change ; there the Minister of the Finances tells him such a thing cannot be, or, in other words, cannot be paid for."

The official courtesy and real kindness shown to Murchison in the metropolis made the leave-taking more than a matter of mere form. From one and all of his friends he received the heartiest congratulations and good wishes, with the expression of a hope for his speedy return. He notes, for instance, that Count Cancrine, the virtual Prime Minister of the Empire, " embraced us, kissing me three times ; and thus encouraged with every promise if I would return, we took our leave."

In spite of fogs and other delays, including a feverish attack, the result of the last week of excitement and conviviality in St. Petersburg, our traveller reached the mouth of the Humber on the 1st November. The last record in the Russian journal, written while the vessel was within a few miles of the Yorkshire coast, is as follows :—

" Seven months and seven days have now elapsed since I left my home on a fine day in the end of March, and I hail Old England with a shining sun again after having travelled through space equal to the diameter of the earth. The Kirghis, the Kalmuck, and the Bashkir excitements are now to give way to plain English comforts, of which I have neither tasted nor thought since I bade adieu to them."

Thus ended Murchison's Russian campaign. The ample record which is given in the great work by his colleagues and himself has made the general scientific results long familiar to geologists. The geological structure of the Russian provinces was now for the first time broadly sketched out and mapped so as to bring the rocks of one half of the European continent into family relationship with those of the other half. Nor were the benefits conferred only on the country in which the long and arduous journey had been made. New light was thrown on questions of general geological import, such as the structure of mountains, the physical geography of the times of the Old Red Sandstone, the classification of the Devonian and Old Red Sandstone rocks of Western Europe, the history of the earlier part of the Carboniferous period, the true order and relations of the red rocks lying between the Coal-measures and the base of the Jurassic series, the former extension of that ancient sea of which the modern Caspian and Sea of Aral are but the diminishing fragments, the southern extension of the ice-borne boulders carried during the Ice Age from Finland and the north far into the low plains of Europe, the occurrence of gold and its distribution in the old alluvia of rivers. The campaign indeed proved to be most fruitful in its issues. It raised Murchison to the same place with regard to the geology of Russia that Pallas fills in its botany.[1] It opened out a new field for research, and paved the way for the good work which has since been done in Russia by other and later observers.

On Murchison himself its influence was profound. It gave breadth to his method of dealing with palæozoic rocks; it increased his aptitude in applying the evidence of fossils to determine questions of geological chronology, and it strengthened his confidence in his Silurian and Devonian work, and in the principles on which that work had been based. Bringing him too into constant and intimate association with foreigners and foreign ways of life and thought, the Russian campaign increased in a high degree

---

[1] Helmersen, *Bulletin de l'Acad. Imp. de St. Petersbourg*, tom. xvii. 1871, p. 295 *et seq.*

his sympathy and respect for men and things abroad, removed from him much, if not all, of that insularity of feeling of which his countrymen are so often accused, and made him more than ever the considerate friend and courteous host of all scientific brethren whose lot brought them to this country, no matter from what quarter of the globe they might come.

Whether the influences of this bold and skilfully conducted journey were altogether beneficial may be matter for doubt. In the course of a few months the geological structure of a vast empire embracing the greater part of Europe had been sketched out—a feat to which there had probably been no parallel in the annals of geological exploration. The success of the campaign and the applause which that success brought from all quarters, were so great that a more than usually well-balanced nature might well have felt the strain too severe to keep its equipoise. From this time forward characteristics which may be traced in the foregoing narrative became more strongly developed in Murchison's character. In his letters and in his published writings his own labours fill a larger and larger space. His friends could trace an increasing impatience of opposition or contradiction in scientific matters, a growing tendency to discover in the work of other fellow-labourers a want of due recognition on their part of what had been done by him, a habit, which became more and more confirmed, of speaking of the researches of his contemporaries, specially of younger men, in a sort of patronizing or condescending way. He had hitherto been, as it were, one of the captains of a regiment; he now felt himself entitled to assume the authority of a general of division. To many men who did not know him, or who knew him only slightly, this tendency assumed an air of arrogance, and was resented as an unwarranted assumption of superiority.   But they who knew Murchison well, and had occasion to see him in many different lights, will doubtless admit that these failings were in large measure those of manner, and at the most lay openly on the surface of his character.   You saw some of them at once, almost before you saw anything else.   Hence it was natural enough that casual intercourse with him should give the impression of a man altogether wrapt up in his own work and fame.   Yet underneath those outer and rather forbidding peculiarities lay a generous and sympathetic nature which inspired many an act of unsolicited and unexpected kindness, and which was known to refuse to be alienated even after the deepest ingratitude.   The success of the Russian researches probably quickened into undue prominence some of the less pleasing features in Murchison's character, but they in no way lessened the measure of kindly interest and sympathy which, in spite of the way he often chose to show them, were those of a true friend.

CHAPTER XVI.

WITH the prestige which the Russian geological tour had given him all over Europe, Murchison returned to resume his town life in London. There lay a vast amount of work before him to be done this winter (1841-2). First of all the notes of the explorations in Russia had to be carefully worked out in anticipation of the visit which it had been arranged should be paid to him by his fellow-travellers, with the view of settling their plans for the preparation of their conjoint volumes on the geology of the Muscovite dominions. The experience which the writing of the *Silurian System* had furnished warned him that his new literary venture would be no easy task; we shall find, indeed, that just as in the case of the growth of that work, so in the elaboration of *Russia and the Ural Mountains*, the progress of his pen, slow enough of itself, needed to be continually sustained by fresh arguments with the hammer. Only now, the intervals of field-work, instead of taking the geologist to old haunts, social and scientific, in Wales and the Border counties, led him to wide digressions into Scandinavia, France, Germany, Poland, Russia—in short, into many far separated tracts of the Continent, whence fresh evidence could be gathered bearing on what had come to be his great geological quest—the true order and classification of the older fossiliferous rocks of Europe.

But besides this main piece of work, he had now to take his place and perform personally the duties of President of the Geological Society, an office to which, as we have seen, he had been for the second time elected, just before he started on his second journey to Russia.   Since he had previously filled the chair he had vastly increased his reputation.   Moreover, the fortune inherited by Mrs. Murchison had very considerably augmented his income ; hence, while eager to sustain his position with dignity and hospitality, he found himself much more able to do so on a large scale than in the old and more modest days at Bryanston Place.

Add to these various avocations the numerous and exacting calls upon the time and thought of a man who occupies a prominent place in London society—calls which, though now increasing enormously on Murchison's hands, he yet strove to meet as far as he could—and we see what the change must have been from the wilds of the Urals to the turmoil of London.

The narrative now to be followed will lead us through the doings of the busy years which culminated in the publication of the work on Russia.   It was during that time that the classification developed in the *Silurian System* received its broad basis in Europe.   In that time, too, the seeds began to germinate of the estrangement which utterly destroyed the ancient brotherly friendship between Sedgwick and Murchison.   There is thus a special interest attaching to this period in relation both to Murchison's life and to the progress of palæozoic geology.

The following letter takes us at once into the midst of the work and play of the winter :—

"16 BELGRAVE SQUARE, *January 25th*, 1842.

"DEAR EGERTON,—My ancient sympathies are not so entirely destroyed that I do not feel for your loss of twenty-five couple of good hounds! and the only compensation is, that we have a chance of seeing more of yourself. Humboldt declines the proposed festival, thanking me for the offer of this 'noble mark of English kindness,' but as the King stays only eight or nine days, and has nine thousand things to do, the thing was impracticable.[1]   Last week I was at Beaudesert trying to shoot in snow, but not prevented during two days from geologizing the fine high wilds of Cannock Chase among the old Marquis's blackcocks, grouse, and big boulder-stones. Then I went to Lord Dartmouth's, where I met a large party and read an inaugural address to the Midland Geological Society, and made five speeches after dinner (Lord Ward in the chair) to all the ironmasters, the most effective hit being when, in the absence of other fighting men, I stood up for the army and navy, and talked of a withered laurel or two which I picked up under the 'Old Duke.'   That name was a talisman among good loyal folks like the Dudleyites.

"I shall see Humboldt, I hope, *chez moi* one of these days, but the devil is that I am losing the best shooting of the year.   I shall read all my discourse [2] this year at the *morning meeting*, so that we may have a real jollification at the Crown and Anchor, after which I fear I shall scarcely be able to face the Earl's symposium."

[1] The King of Prussia was then on a visit to England, with Humboldt as one of his suite.

[2] The President's address at the anniversary of the Geological Society in February.

Before the end of the year the inaugural address mentioned in this letter had been printed and circulated among his friends. From one of these, the facetious Sydney Smith, he received the subjoined acknowledgment :—

" DEAR MURCHISON,—Many thanks for your yellow book, which is just come down to me. You have gained great fame, and I am very glad of it; had it been in theology, I should have been your rival, and probably have been jealous of you, but as it is in geology, my benevolence and real good-will towards you have fair play.

" I shall read you out loud to-day. Heaven send I may understand you : not that I suspect your perspicuity, but that my knowledge of your science is too slender for that advantage—a knowledge which just enables me to distinguish between the Caseous and the Cretaceous formations, or, as the vulgar have it, to know chalk from cheese.

" There are no people here, and no events, so I have no news to tell you, except that in this mild climate my orange-trees are now out of doors, and in full bearing. Immediately before my windows, there are twelve large oranges on one tree. The trees themselves are not correctly the Linnean orange-tree, but what are popularly called the bay tree, in large green boxes of the most correct shape, and the oranges well secured with the best pack-thread. They are universally admired, and, upon the whole, considered finer than the Ludovican orange-trees of Versailles. Best regards to Mrs. M.—Yours, my dear Murchison, very truly,

" SYDNEY SMITH.

"TAUNTON, *December* 26, 1841."

Two other letters of the same correspondent, called forth

by similar presents of copies of Murchison's memoirs and addresses, may be given here :—

"DEAR MURCHISON,—Many thanks for your kind recollection of me in sending me your pamphlet, which I shall read with all attention and care. My observation has necessarily been so much fixed on missions of another description, that I am hardly reconciled to zealots going out with voltaic batteries and crucibles for the conversion of mankind, and baptizing their fellow-creatures with the mineral acids ; but I will endeavour to admire and believe in you.[1]

" My real alarm for you is, that by some late decisions of the magistrates, you come under the legal definition of *Strollers*, and nothing could give me more pain than to see any of the Sections upon the Mill calculating the resistance of the air, and showing the additional quantity of flour which might be ground *in vacuo*—each man in the meantime imagining himself a Galileo. We have had Mrs. Grote here : Grotius would not come. The basis of her character is rural, and she was intended for a country clergyman's wife ; but for whatever she was intended, she is an extraordinary clever woman, and we all liked her very much.

" Mrs. Sydney has eight distinct illnesses, and I have nine. We take something every hour, and pass the mixture from one to the other, as Mrs. M. and you do the bottle.

" About forty years ago I stopped an infant in Lord Breadalbane's ground, and patted his face ; the nurse said, ' Hold up your head, Lord Glenorchy.' This was the President of your Society ; he seems to be acting an honourable and enlightened part in life ; pray present my respects to

---

[1] Reference is here made to the proceedings of the British Association. Lord Breadalbane was President in 1840.

him and his beautiful Countess.—Yours, my dear Murchison,
very truly,   SYDNEY SMITH."

"DEAR MURCHISON,—Many thanks for your address,
which I shall diligently read. May there not be some one
among the infinite worlds where men and women are all
made of stone—perhaps of Parian marble? How infinitely
superior to flesh and blood! and what a paradise for you, to
pass eternity with a Greywacke Woman !!!—Ever yours,
                                     " SYDNEY SMITH."

The anniversary address given to the Geological Society
in February 1842 was a laboured production, occupying
forty of the closely printed pages of the Society's *Proceedings*, and must have somewhat exhausted both reader
and audience from its mere length. During the interval
of ten years which had passed away since Murchison
read a similar discourse, his favourite science had in
some departments made rapid strides; but in none had
its progress been so remarkable as in the classification of
the older fossiliferous rocks, a result which sprang in great
measure out of his own labours. Naturally therefore he
dwells upon his share in the triumphal progress of geology. Giving his brethren of the hammer a sketch of
the steps by which the classification had been worked
out, he alludes to his adoption of the term "Silurian," remarking that he had some pride in restoring that name to
currency in remembrance of the boast of the Roman general
Ostorius, who, on conquering Caractacus, declared that he
had blotted out the very name of the British Silures from
the face of the earth. He justifies the use of a geographical
terminology, and very pointedly calls attention to the

absence of any zoological boundary between the Cambrian
and Silurian systems, a fact which had already been admitted by Sedgwick.[1]   He gathers together with manifest
satisfaction the evidence of the extension of the Silurian
system in Europe, Africa, America, Australia, and the South
Seas.   The Geological Survey had been making progress in
South Wales, and had begun to grapple with the problem
as to the separation between Cambria and Siluria.   While
alluding to its progress under the leadership of De la Beche,
Murchison refers again to the work of the Survey in Devonshire, and to his own labours there and on the Continent in
conjunction with Sedgwick.   The rocks of Devonshire lead
him to say a few kindly words of Hugh Miller's *Old Red
Sandstone*, which had recently appeared, and to speak of
the wonderful series of bone-cased uncouth fishes furnished
by the Old Red Sandstone of Scotland and Russia.   Among
his allusions to fossils there occurs a reference to the remarkable announcement by Ehrenberg of the occurrence of
still living species in the Cretaceous rocks, a fact which
showed " the danger of as yet attempting to establish a

[1] *Proc. Geol. Soc.*, iii. 641.   The principle on which Murchison had
proceeded in his Silurian classification was that which had guided William Smith among the Secondary rocks—"Strata identified by their
organic remains."   If, therefore, he found a series of strata containing
nothing but Silurian fossils, he was logically bound to class it as Silurian.
This was the inevitable step in store for him, and that he saw it coming
seems to be indicated in this address.   He says that "the term 'Cambrian' must cease to be used in *zoological* classification, it being in that
sense synonymous with 'Lower Silurian,'" and adds, that the line of division placed on his map between the two series has no longer any palæontological significance.   He hints that the Cambrian series is but a local
subdivision of the same great palæozoic group.   Sedgwick's susceptibilities do not seem to have been roused at this time, but the subsequent perusal of this address and that for the next year led him to protest
against the proposal to wipe out the Cambrian system from geological
nomenclature.   See Sedgwick's *Letters to Wordsworth*, Letter V. p. 86,
and *postea*, p. 380, *note*.

nomenclature founded solely on the fauna and flora of former conditions of the planet." After eulogies of foreign geologists, and notably of L. von Buch, to whom he conveyed the Society's Wollaston medal, he winds up his oration with a long disquisition on the glacial theories which had been discussed at Glasgow, and regarding which he had then announced his intention " to show fight." He refuses to allow Agassiz to cover the northern parts of our hemisphere with sheets of ice, but admits that the evidence compels him to concede that the land was submerged beneath an ocean over which ice-rafts and icebergs sailed southwards.

Here is Murchison's own report of his discourse and the meeting, as sent at the time to Sedgwick :—

26th February 1842.—" The anniversary went off gloriously, though I say so. The morning discourse was well received, and in truth I put a deal of powder and shot into it, foreign and domestic, and took so much pains as to stop my original work on Russia. . . . [I write] as well as a man can whose first soiree begins to-night with probably 200 or 300 people coming ! The morning room was full, and I read for two hours without losing a man. I entered at length into the Silurian and ' Palæozoic' question. . . . I defended the temporary division set up between your lower slaty rocks and my superior groups on the ground of positive observation of infraposition, and if in the end (as I now firmly believe) no suite of organic remains will be found, even in the lowest depths, which differs on the whole from the Silurian types, why then we prove the curious law that in the earliest inhabited seas of our planet the same forms were long continued.

" I took care to show that any other plan than that

which we adopted would have led to fatal errors, such as
' Système Hercynien' and other hypotheses, and that now
all must come right, to whatever extent (and the extent can
probably never be defined) the base of the Lower Silurian
zoological type may be extended. . . .

"Our dinner went off '*con amore*,' and every one says
it was the best (Adam Sedgwick only wanted) which we
ever had.   I did my best to make it of a public character,
and had my two Knights of the Garter, one on either side
the President, and the representative of my Emperor
Nicholas.   Brunnow spoke admirably, and I never heard
Lord Lansdowne speak so well as for the toast of 'The
Universities of this Land.' . . . Having no science to go
to and snore over at night, the *cœna et nox* went off just as
I could have wished it, and I so handicapped my running
horses that they each made play where I wanted it.   I
send you a scrap from the *Morning Post*, possibly written
by—— . . . Knowing that he was going to furnish something, I popped my speech [about the Emperor and Baron
Brunnow] into his hands, being well aware that words are
weighed at St. Petersburg.   Tell Whewell of our frolics."

Among the survivors of that small band of enthusiasts
who founded the Geological Society, one of the most prominent still took, even in his old age, a keen interest in the
Society's affairs.   No face was more familiar at the meetings than that of G. B. Greenough, no voice more often
heard in the discussions.   Every new theory, or proposed
reform of an old one, every suggested change in the established nomenclature of geology, was sure to receive keen
scrutiny, and probably more or less of active or at least
passive opposition, from the veteran President of the Society.

He used even to astonish the propounder of some novelty
by demonstrating, or at least endeavouring to demonstrate,
that what was thought to be new was really only another
version of what had been known long before, had perhaps
been even taught by Werner himself. We have seen that
this happened to be his mood of opposition when the
Devonian question came up for discussion before the Society.
And yet with this adherence to his early habits of thought,
and with a doggedness of opposition which, though always
courteous and good-natured, must often have been provoking
enough, Greenough retained the deep respect and esteem of
every member of the Society. This was manifested now
by a movement to perpetuate his features in a bust, to be
placed and preserved in the apartments at Somerset House.[1]
Murchison took a leading share in the organization of this
scheme, which when propounded to Greenough drew from him
the following acknowledgment, addressed to Murchison :—

March 30, 1842.—"For the exertions I have made in
behalf of the Geological Society I have been most liberally
remunerated by the confidence reposed in me at all times
by the body at large, and by the invaluable friendships
which I have formed with many of the members. I accept,
however, with much pleasure, the distinction now presented
to me, viewing it, as I do, not merely as an acknowledgment
that I have faithfully discharged my duty, but also as a
stimulant to exertion in others, and above all as a guaran-
tee that those principles which, in the infancy of our establishment, were resolutely insisted upon as essential to
the well-being of every scientific institution, will continue
to be cherished in the Geological Society, not only in the

[1] It was intrusted to Westmacott.

lifetime of its founders, but long after their decease.—
Yours sincerely,   G. B. GREENOUGH."

Whilst the geologists of Britain were in this graceful
way crowning with honour the latter days of one of their
earliest fellow-workers, another member of the brotherhood of hammerers was about to begin a career which has
gained for him a high place in the annals of geological discovery, and with both of these events Murchison was
intimately associated.   The Provincial Legislature of Canada
had voted a sum of £1500 for a geological survey of the
province.   With the view of securing a competent person
to undertake the duties of such a survey, the Governor-
General applied to the Home Government, mentioning in
particular the name of Mr. W. E. Logan, and requesting
Lord Stanley to ascertain whether, in the opinion of the
Geological Society of London, or other competent authorities, he was considered to be qualified.   This official request
was communicated to Murchison, as President of the Society.
Mr. Logan had already distinguished himself by some
admirable surveys of the South Welsh coal-fields, and by
observations on the formation of coal.   He had worked
enthusiastically as a volunteer in De la Beche's staff of the
Geological Survey, and his large sections, drawn to a true
scale of six inches to a mile, led to all the subsequent admirable sections by De la Beche and his colleagues.   Murchison,
who knew these labours well, and had made use of them in
his Silurian map, recommended the proposed appointment in
the warmest terms, adding that it would " render essential
service to Canada, and materially favour the advancement
of geological inquiry."   Shortly afterwards Mr. Logan re-

ceived the appointment, and returned to Canada, his native country, to lay the foundations, and for about thirty years, in spite of many discouragements, to work out the development of one of the most important and successful geological surveys that have ever been carried on in any country.

Summer had brought back leaf and blossom ere bags and hammers were furbished up anew for field-work. A plan which had been discussed the previous year in Russia was now to be put into execution, viz., that Murchison should with his comrades make a careful examination of some of the best sections of the older rocks of Britain, for the sake of renewed and more definite comparison with those of the Continent, and especially of Russia. Count Von Keyserling duly arrived, and after the usual and indispensable hospitalities in London, Murchison and he started on their English tour. Beginning with the Isle of Wight, they first worked their way over the Secondary formations westward as far as Cheltenham and the Malverns. Then they turned northwards into the old Silurian region, lingering at the rocks and country-houses which had been Murchison's favourite haunts ten years before, and passing across the undefined and increasingly indefinable line between Cambria and Siluria, away over Sedgwick's domains even to the far promontories of North Wales. Turning still northwards, the two geologists halted in Durham to compare the rocks and fossils of that county with those of the Russian province whence the term 'Permian' had been taken. The northern coal-fields, so like in some respects to those of Russia, offered many points of interest for comparison. So intent, however, were the travellers in gathering materials for the illustration of their Russian work, that they pro-

longed their journey into Scotland, tracing the red sandstones which emerge from under the coal-bearing tracts, and
in which they saw much to remind them of the great areas
of Old Red Sandstone in Russia. Crossing to Carlisle on
their southward journey, they worked their way through the
Lake district, thence down the great Carboniferous Limestone
tracts of Yorkshire and Derbyshire into the Staffordshire
coal-field until they once more found themselves on the
slopes of the Malverns.

Such was the round of country examined. One or two
parts of the journey deserve notice from the sequel to which
they led. In the course of their traverse from the Silurian
into the Cambrian region, the travellers were as unable
as anybody had ever yet been to draw any satisfactory line
between the two tracts. Mineralogically there was really no
true boundary line, and zoologically it had been agreed even
by Sedgwick himself that no distinct assemblage of fossils
had been ascertained to belong to the Cambrian series.

The Geological Survey under De la Beche had now been
extended into Wales. When Murchison and Von Keyserling
were on their tour, the Survey forces were at work among
the Silurian and Cambrian strata, and had already, after
much careful mapping, made out some important points
regarding the relation of these strata. Some of these are
referred to in the following extracts from a letter by De la
Beche to Murchison. *Llandovery, 31st July* 1842.—"Touching the Silurian system, heaven knows where it is to end
northwards in this land! it goes in great rolls, and *no mistake*, a long way beyond the Caermarthen (Ordnance map)
sheet. No want of fossils; in fact, organics and sections
all going to prove the same thing. The cleavage no doubt

is abominable, but by *very* careful hunting of all the natural sections, and giving lots of time to it, the affair has at last come out clear enough. . . . It would be a long story to go further into the *old story* hereabouts; that your Silurian system must have a jolly extension at our hands over the rocks of this land seems certain."

The extension referred to was mainly due to the labours of Mr. Ramsay, who, since he left for Tenby, had been hard at work among the Welsh rocks.   On the 7th August of this same year he reported progress to Murchison as follows: " I have gradually gone over the whole of the *ci-devant* Cambrians between St. David's and Llandovery, and I can clearly show, particularly since I came here [Pumsant], that all *your* rocks, under a somewhat different form, spread over the surface of the land at least as far as Cardigan. . . . I should much like to show you some of the evidences of this Cambrian revolution."

These were important labours in the progress of British geology ; but their special interest in the present narrative lies in their relation to Murchison and his views.   It will be seen that they confirmed his belief in the extension of the Silurian forms of life among the older rocks, and they no doubt contributed not a little to foster that spirit of confident assertion which marked his next oration to the Geological Society.   He counted as personal friends the men by whom these researches had been conducted, but until this summer, when he took Count Von Keyserling with him, he had not become acquainted with the way in which their actual work in the Geological Survey was carried on. Phillips was then busy " running a section " across the Malverns.   So Murchison and his Russian companion went round to see. They found their friend, on a bright September morning, on the summit of the Beacon, busy with his theodolite, and learnt something of the laborious detail of geological surveying, so different from the hop-step-and-jump kind of work with which their Russian experiences had familiarized them.

An important change took place this autumn in the Geological Society. Lonsdale, feeling the growing weakness of his health, and the increasing urgency of the calls of the Society upon his powers, had resigned his Curatorship, with the purpose of seeking rest in retirement. As Murchison had been the means of bringing him to London, and had enjoyed his close friendship, as well as the quite invaluable aid which Lonsdale cheerfully rendered in palæontological and other matters, he now took an active part in promoting the subscription for a testimonial to the worthy Curator, expressive of the universal regret at his retirement. A silver cup, together with a sum of £600, were presented by Murchison and Fitton, in name of the subscribers, to Lonsdale, who, unable at the time to find a vent for his feelings, sent a characteristically modest and grateful note to Murchison. "Should life be granted me," he said, " I purpose to pursue the study of fossil polyparies, and it will be a source of personal gratification if my friends will transmit to me any specimens they may think me capable of examining, and for the means of conducting this inquiry I shall be indebted to them."

For fourteen years Lonsdale had been in the midst of all the activity of the Geological Society. During that time not a publication had been issued by the Society which did not owe much to his careful supervision. But the official work which he performed so well, and which undoubtedly had no small influence on the general progress of geology in England, represented only a part, and perhaps not even the chief part, of the obligations under which he placed the members of the Society.   There were few of the geologists engaged, like Murchison, in active research and in independent publication, who had not recourse to Lonsdale as an ever ready and sagacious helper.   In a body of men who, busy with the same pursuits, are always necessarily to some extent rivals, there must needs arise ever and anon occasions when unwarranted assertions on one side are met by more or less angry recrimination on the other, and when the truth of the question in dispute becomes clouded by the personalities of the disputants.   Such cases, despite the glowing eulogiums in presidential addresses, were not unknown in the Geological Society.   Lonsdale's perfect impartiality and candour, and his tact and shrewd sense, enabled him to moderate these ebullitions, and to preserve the harmony of the brotherhood.

Though he now retired from Somerset House, he could not so easily wean himself from the Society and the pursuits of its members, with whom he had been so long and so intimately associated.   He went down to Dartmouth to enjoy pure air and give himself up to the unremitting study of his favourite branch of inquiry, the structure of fossil corals.   But we find him carrying on still, as of old, a voluminous correspondence with the President on affairs of finance, the preparation of the Society's Transactions, the choice of office-bearers, and other matters of business, besides the more strictly scientific subjects on which they were both engaged.

Lonsdale's resignation brought into the service of the

Society, and prominently into geological pursuits, another
naturalist of greater knowledge and wider fame.   When the
Curator's determination to leave came to be known, various
names were talked about in reference to the supplying
of his vacant post, among them that of Hugh Miller.
But, after some delay, the final decision among nine can-
didates was made in favour of Edward Forbes, who had
recently been chosen Professor of Botany in King's College, and whose brilliant researches in the Ægean gave
promise of a distinguished career as a naturalist and palæontologist.

The appointment of Forbes to be Curator of the Geological Society must be regarded as an event of considerable
importance in the history of geological progress in Britain.
While still an enthusiastic student of natural history under
Jameson at Edinburgh, he had struck out into that little-
trodden path of research in zoological and botanical distribution wherein he continued to be throughout his too short
life the great pioneer.   Already, by excursions in this
country, in Scandinavia, and in Switzerland, he had been
led to recognise the connexion between geological changes
and the present grouping of plants and animals.   Fortunately provided with further and more advantageous
opportunities of concentrated research, by being attached to
Captain Graves's surveying ship in the Ægean Sea, he had
thrown quite a fresh light on the way in which the pro-
secution of zoological research might be made subservient
to the elucidation of some of the most interesting questions
in geology, such as the history of existing species of animals
and the geographical changes of which they have been the
witnesses.   By these bold and original investigations he

had in a special manner attracted the notice of geologists.[1] And now that his duties at Somerset House brought him into direct relationship with the leaders of geological inquiry in Britain, his subsequent scientific work took thenceforward a more decidedly geological aspect.

It is not, however, in his relations to the general progress of the science, but in his connexion with the more limited field of palæozoic geology, that the advent and work of Edward Forbes require notice here. His position as Curator at Somerset House undoubtedly led directly to his subsequent appointment as naturalist to the Geological Survey,[2] to the admirable arrangement of the palæontological collections placed under his charge in the Jermyn Street Museum, and to the good service which he rendered in working out the natural history of Silurian and Tertiary rocks. It brought him also into intimate personal relations with Murchison, De la Beche, Ramsay, and the others on whom the progress of palæozoic geology in this country mainly depended.

The winter of 1842-3 was with Murchison a very busy one. It was to be his last season of office as President of the geologists, and besides the proper official duties, which he conscientiously discharged, he entered with renewed zest into the social festivities for which the Belgrave Square mansion had now become well known. There were few men of note in literature, politics, science, or art to whom

---

[1] In 1841 he had received from the Geological Society the balance of the Wollaston fund, amounting to £30, to assist him in carrying on his researches.

[2] The actual proposal of Forbes to De la Beche for employment in the Survey was made by Mr. A. C. Ramsay, who had known the young naturalist well since 1840.

the soirees of the President of the Geological Society were not, or might not have been, familiar.

At the anniversary in February, when he would resign office, he had determined to give an address to the Society containing a detailed report of progress, and in particular a more pointed statement of his position with regard to the impending changes in Cambrian and Silurian nomenclature. How he meant to proceed is shown in the subjoined letter of 16th October :—

" MY DEAR SEDGWICK,—On the 1st of next month I go to press with the work on Russia, which with amplifications and emendations is composed of the memoir referred to you last year, and two which I have read since on other parts of Muscovy and on the Ural Mountains. The country is described in ascending order, and I therefore must cast my Silurian chapter at once into type, with a preamble on 'Palæozoic rocks,' which shall render my views intelligible to the Russians, for whom the work is hereafter to be translated. In doing this I necessarily give a little sketch of our own operations in the British Isles and in the Rhenish Provinces, and then go on to show how Russia completes the proofs desired, and confirms our views. Now in effecting this to my satisfaction, I wish to have your own authority to speak out concerning the Cambrian rocks zoologically considered. You know as well as myself that on those parts of the Continent which we have seen together, there is but one type of fossil remains beneath an unquestionable Devonian zone, and that we have called Silurian. The same is still more clearly exhibited in Russia in the limestones, sandstone, and shale, which lie beneath true Old Red Sandstone, filled both with fishes of Scotland and shells of Devon. The Silurian rocks of Russia, Gothland, and Sweden rest at once on the crystalline slates of the north. The same succession has been recently established (zoologically) in Brittany by Verneuil and d'Archiac this summer, though there they have inferior slaty rocks without fossils unconformable to Caradoc sandstone. Whilst these inquiries have been deciding the zoological succession on the Continent, and extending it even into Asia, our own region at home has been silent. I was rejoiced therefore when I knew you had been again into North Wales, and that you had taken young Salter with you, because you could then make up your mind to put your oracle out, without having it trumpeted forth by others.

" In the meantime, besides what Mr. Maclauchlan stated in respect to Pembrokeshire, De la Beche and his workmen assure me, that the whole of that tract_is nothing more than Caradoc sandstone and Llandeilo flag, or Lower Silurian, folded over and over in troughs, and exceedingly altered by intrusive rocks and changed by crystallization and cleavage. They contend also that the very same identical fossils, *in the very same strata* as those which I have described and figured as Lower Silurian at Noeth Grüg, north of Llandovery (and only a few miles from the Old Red escarpment), are repeated over and over, up to the sea-coast at Cardigan, and to the north of it. To this I cannot say nay, because in my work I have described descending passages into what I certainly conceived, without perhaps sufficient examination, to be a great *inferior* slaty mass, and in which I never observed the fossils in question. If their position is true it would be in vain to contend for Cambrian rocks in South Wales, and certainly not as identified by organic

remains, though I am certain *there are* inferior slaty grauwackes at St. David's, like those of the Longmynd in Salop,
which are much older than my fossil Silurian—and of this
you know I have decisive proofs in Salop, where the Caradoc sandstone rests on the edges of the Longmynd.[1]

"But the question is, If there are no rocks containing
fossils differing from those published as Lower Silurian in
South Wales, are there such in North Wales, where limestones appear in the oldest slaty masses, and the whole
is expanded and broken up by the anticlinals you have so
well described? As to Bala, you know that its examination
will do nothing in establishing a distinction, and fortunately I have said so very distinctly in my *Silurian System*,
and have asked the question, To what extent will the
Orthidæ and *Leptænæ* in question be found to descend into
the Cambrian rocks, and if they really constitute the Protozoic type? (p. 308, *Sil. Syst.*)

" I mention this now because I understand from Lonsdale
that Mr. Sharpe is going to read a paper at the second
meeting of the Geological Society, in which he is to show
that the Bala limestone is nothing more than a calcareous
course in the middle of the Caradoc sandstone.   I do not
see how he is to do this *stratigraphically*, but as I never
made the transverse section but once, and in your company,
I do not pretend to be armed with sufficient proofs that the
limestone is inferior to the slaty flagstones on the eastern
side of the mountain in which *Asaphus Buchii* and Silurian

----

[1] This happened to be a blunder on Murchison's part ; he was right as
regarded the unconformability, but wrong in the position which he had
assigned in the *Silurian System* to the overlying strata.   These are what
we now term Upper Llandovery (that is, at the base of the Upper Silurian
series), and not Caradoc.

Orthidæ occur; and on this point, by way of parenthesis, I should like to be furnished with your view, in order that I may keep the 'Sharp' fellow in his place, should he transgress bounds.

" But to come to the question : If Bala is zoologically Lower Silurian (and that you have yourself now stated in your Letters to Wordsworth), if Coniston Water Head and Ambleside (at the latter place Keyserling and myself convinced ourselves of the same) is the same thing, and if no older rock is known to contain fossils in Cumberland, it follows, that the only fossil type which remains to be appealed to is that of the Snowdon slates. In our recent visit, Keyserling and myself collected a good many fossils both on the north and on the west flanks of that mountain, and my friend, who is a very good conchologist, came to the conclusion on the spot, that the prevalent and abundant forms are two or three species of *Orthis* (*flabellulum* and *alternata*) well known in Lower Silurian and Caradoc, with a rare new form of *Leptœna ;* and Sowerby, who has since seen our lot, writes to me to the same effect, and tells me that Salter's determinations with you came to the same results.

" Now, I have no intention whatever of writing upon this point, except in my exordium on Palæozoics touching Russia, where I have to treat of them over an area as large as all our Europe together. On that occasion, and also in taking leave of the geologists on the 17th February, I *must* deliver my opinion. Your Wordsworth letter is before me, and is a meet subject for my comment, but I wish to have something from you touching North Wales. If this is not done, De la Beche and Co., advancing from South Wales, will

have the credit with the public of correcting you. But if
you now say that the slaty region to the north-west of the
Silurian rocks was left undefined as to *fossils,* on account
of your never having examined the forms you so long ago
collected (and take any line you please, either to contend or
not for great thickness of the lowest fossiliferous strata),
then I shall be at ease, and know how to use your authority
as well as my own.[1]

[1] Murchison's anxiety to carry Sedgwick with him, if possible, in his
change of the Silurian base-line, is well shown in this letter and in the
following postscript to it :—" In the part which specially refers to what I
have been writing to you about, I should, in case you will authorize me,
propose to write something such as follows :—After asking ' if no efforts
had been recently made to determine the point if there were or not a
group of older fossils than the Lower Silurian, and some paragraphs
relating thereto,' I go on to say, ' Judging from their infraposition, great
thickness, and distinct lithological characters, it was presumed (when the
Cambrian system was so named) that these greatly developed inferior
slaty rocks would be found to contain a class of organic remains peculiar
to themselves, the more so as the few forms then discovered in them
seemed to differ from the Lower Silurian types. Subsequent researches
have, however, decided otherwise. In the slaty region of the north-
west of England, of which by hard labours he so long ago rendered
himself the master, Professor Sedgwick has now satisfied himself that
the lowest organic remains which can be traced are no others than those
published as Lower Silurian, whilst in revisiting the mountains of Cambria and Snowdon, whose framework he was the first to explain, he has
come to similar conclusions respecting the oldest fossiliferous tracts of
North Wales.'
" ' In the meantime, through the labours of the Ordnance Survey,' etc.
Then Mr. Sharpe *et hoc genus omne.*
" This is the form in which I should wish to place the case, both
because it is in my mind *quite true,* and also because, as I have said in
my letter, I wish you to speak in your own place."
Sedgwick made no objection at the time to this statement of his views.
On the contrary, when he received the proof-sheets of the address he
made comments on other parts, but, so far as can be judged from the
letters still extant, offered no criticism whatever on the proposed narrative given in the preceding extract. He returned the proofs with the
remarks, " The papers are excellent, and use my hints as you think
right. . . . I have looked over the slips and made marks. . . . I did look over
the peroration. It is very good." It was, to say the least, unfortunate

" The triple zoological division of the Palæozoic rocks (exclusive of the Magnesian Limestone) is now so very generally proved to the very eastern extremities of Europe, that it is well that we who have been the agents in first enunciating it should not be frightened and driven out of our fairly won views because the Cambrian tail-piece *was not finished off.* For my own part, I am as convinced as it is possible to be, that we have now thoroughly ascertained not

that, if he had really any strong objections to the statements in the address, he did not frankly express them at the time when the proof-sheets were sent to him. Had he done so we can hardly believe that he could afterwards have found occasion to say of any sentence in that document : " I smiled when I read this strange passage ; but I did not think it worth while formally to contradict it ; in omission and commission it is a virtual mis-statement of the facts."—(*Letters to Wordsworth,* later edition, p. 87.) Surely by first sending his friend a sketch of what he meant to say, and then the proof-sheets of what he had said, Murchison showed no common care to secure his concurrence. It is hard to understand why Sedgwick should have entered into verbal and other criticisms in the most friendly and even jocular style, and yet have left untouched a passage which raised a " smile," and which he felt to be " a virtual mis-statement of the facts."

But what was the " strange passage " which called forth these sharp words? As quoted and italicised by Sedgwick himself, it ran as follows : " We were both aware that the Bala ·limestone fossils agreed with the Lower Silurian ; but *depending upon Professor Sedgwick's conviction* that there were other and inferior masses, also fossiliferous, we both *clung to the hope* that such strata, when thoroughly explored, would offer a sufficiency of new forms to characterize an inferior system."

On this passage he remarks as follows :—"When the author states ' that we both clung to the hope that the Cambrian groups would offer a sufficiency of new forms to characterize an inferior system,' I can only reply, *that the hope to which he clung* was not derived from anything I had ever said or written ; and that I had not, in 1842 and 1843, the shadow of a hope that any new system of animal life, any group of new forms ' marking an inferior system,' would be found among the Lower Cambrian groups. I had constantly expressed, and repeatedly published, *a directly contrary opinion.*" (The italics are in the original.)

Now it will hardly be believed that Murchison's statement is not only borne out by passages in Sedgwick's letters, but seems actually based upon them. In support of this assertion two extracts may be given. Writing to his friend after his autumnal ramble in Wales in 1842, Sedg-

only the Palæozoic, but, as I ventured long ago to call it, the Protozoic type, and that *that* is no other than the striking orthidian Lower Silurian group, which, first rising up on the flanks of old Caradoc, is extended to any thickness you please to contend for.   In this last respect, however, you must have the fear of De la Beche and his trigonometrical forces before your eyes, who, whilst they give 12,000 or 15,000 feet thickness to the South Welsh coal-field, are cutting down our older rocks at a terrible rate. . . .

"Before I left town I presented £600 to Lonsdale, in a silver vase with a suitable inscription.   Fitton accompanied me, and the poor fellow was *quite overcome*.   The deed however had an excellent effect, for his eyes brightened up in the following days, and he wrote me a most affectionate note, saying ' that he was *now* enabled, even in his retirement, to carry on his studies, and that he would go on with that of the Polypifers."

Among the miscellaneous correspondence of this period which the President of the Geological Society carried on, was one regarding a proposed purchase of the island of Staffa.   It was represented urgently to Murchison that as wick says :—" To my knowledge of the *sections* I added nothing last autumn, but I hoped to make out *distinct fossil groups*, indicating a descending series, and marking the successive descending calcareous *junks*.   But, as I told you, I failed."   The italics in this and the next quotation are underlined in the original.   Again, just before the anniversary in February 1843, in reply to Murchison's request for information (in the letter quoted above in the text), Sedgwick remarks, "In regard to N. Wales you know my general views.   I stated last year (see the abstracts) that on unpacking my Welsh fossils I could not discover any trace of a lower zoological system than that indicated in your Lower Silurian types. I did however *expect* to find certain definite groups indicating a succession in the ascending steps of a vast section (certainly many thousand feet thick), and my hope was last September to prove this point, but I failed utterly, as I told you before, and at present I really know no such definite groups."

the island was likely to come into the market, no more fitting purchaser could be found than the Geological Society of London, and that in the hands of that learned body it would remain as a perpetual monument consecrated to the progress of science. It is needless to say that this project never took shape. There is little sympathy in Britain with any such fanciful notions regarding the acquirement of places of great natural interest by the State or learned societies for the good of the country and in the cause of scientific progress. Fortunately that fairy isle is too small and too barren to warrant the cost of protecting walls and notices to trespassers, and its wonders are of too solid and enduring a nature to be liable to effacement by the ruthless curiosity of the British tourist. And so it stands amid the lone sea, open to all comers, lifting its little carpet of bright green above the waves which have tunnelled its pillared cliffs, and which are ceaselessly destroying and renewing the beauty of the sculpture they have revealed.

From the foregoing letter to Sedgwick it is clear that the preparation of the address to the Geological Society, and in particular the forcible enunciation in it of his views regarding the classification of the older rocks, engaged much of Murchison's attention during the winter. When at last the anniversary came he produced a most voluminous oration, extending over eighty-seven closely printed octavo pages, and discussing not only the question lying at the time nearest his own heart, but the general march of geology all over the world. Again he presents to foreign geologists—Élie de Beaumont and Dufrénoy—the Wollaston medal with due laudation. After a kindly and appreciative eulogy of Lonsdale and welcome of Forbes, he plunges at

once into the palæozoic rocks, and is soon in the midst of
Silurian and Cambrian nomenclature, laying down with renewed emphasis the view that his own Silurian deposits
contained the records of the earliest type or *facies* of
organized existence.   In the early summer of the previous
year Sedgwick had written his now well-known letters to
Wordsworth on the Geology of the Lake District, in which he
summarized in popular but accurate form the results of his
long labours among these mountains.   Another observer, Mr.
Daniel Sharpe, already referred to, had been at work upon
the Cumbrian tracts, and transferring his knowledge of them
to the investigation of North Wales, had announced his
belief that Sedgwick's Bala rocks were really, both by fossils
and physical continuity, the very same as some of Murchison's Lower Silurian series.[1]   Sedgwick himself had spent

---

[1] In the beginning of his paper Mr. Sharpe stated that the view of the
infraposition of the so-called Cambrian rocks of Sedgwick to the Lower
Silurian of Murchison was adopted by the latter geologist on the authority of the former.   In long subsequent years, Sedgwick bitterly com-
plained that this was a mis-statement, which Murchison never corrected,
but, on the contrary, proceeded to profit by, though he had abundant
opportunity of rectifying it in this address.   And the inference drawn is,
that Murchison was guilty of disingenuous conduct unworthy of a gentleman, still more of a friend (*Introduction to British Palæozoic Fossils*, p.
lxxiii.)   But, so far from regarding it as a mis-statement, Murchison himself repeats it in this very address.   He says that he steadily relied on
Sedgwick's original opinion, that great masses of the slaty rocks of North
Wales lay below the Silurian rocks.   His respect for Sedgwick's opinion
was profound, and that opinion he believed to have been all along in
favour of the infraposition of all the so-called Cambrian rocks.   This
belief, as we have already seen (*ante*, p. 225, *note*), was commonly held by
geologists, and, if a mistake, Sedgwick never did anything to set it right
until he found some of his Cambrian formations claimed as Silurian, when
he maintained that he had never made any error in his work, except in being
misled by his friend.   The charge of unfair conduct on Murchison's part
was utterly unfounded.   Nothing could have been more candid than the
way in which he acted in this matter.   Equally groundless was the accusa-
tion that he had "stolen a march" upon Sedgwick, unless we are to be
told that under such conduct we must include making our victim privy

part of the summer of 1842 in re-examining some portions of the North Welsh area, with the view of clearing up the difficulties in the way of reconciling his own
work with that of his friend.   But he could not establish
any distinction by means of fossils between the rocks which
he had called Cambrian and those which Murchison had
termed Lower Silurian.   He intimated this to the President,[1]
who now, with evident satisfaction, announces it as further
proof that the Silurian type of organic remains had been
firmly established as the oldest in the geological record.
Murchison further dwells on the important aid given to
his interpretation by the labours of the Geological Survey,
which, as we have seen, had now been extended into the
Silurian tracts of South Wales.   While eulogizing the work
of the Ordnance Geological Surveyors in Wales, he turns to
that of their fellow-labourers, and notably Captain (afterwards General) Portlock, in Ireland, adding words of praise
to his notice of the geological map of Ireland by Mr. (now
Sir Richard) Griffith—that wonderful achievement, which
gives its courageous and undaunted author so honourable
a rank among the great geological map-makers of this
century.

We need not follow the address through its review of
contemporary foreign geology, with its elaborate analysis of
what had then been recently accomplished in Russia, the
Caucasus, Asia Minor, Turkey, the Alps, Hindustan, Aff-
ghanistan, China, Egypt, and North America, or through its

beforehand to the theft, and submitting for his approval the plan by
which he is to be cozened.   Yet Sedgwick asserted that the first intima-
tion he had of Murchison's claim over the Upper Cambrian rocks
as Lower Silurian was obtained accidentally, some years after the seizure
had been made!   [1] See p. 382, *note.*

details regarding the progress of dynamical and palæonto-logical geology. Its main interest for us lies in its relation to the controversy, now imminent, regarding the palæozoic nomenclature and to Murchison's position in that con-troversy. Writing of it many years afterwards he thus expressed himself : " That address embodied all my matured views on the classification of the older rocks, and par-ticularly as to the unity of the Silurian system and the im-possibility of manufacturing a fossiliferous Cambrian system separate from the well-recognised Lower Silurian types. Von Buch, Humboldt, and all the foreign geologists, as well as my colleagues in the work in Russia, saw the necessity of this. I therefore openly proclaimed my conviction that the masses of hard and slaty rocks of Wales to the west of my Silurian map and sections, and which were supposed to be Cambrian, before their order and contents were elaborated by the surveyors and Sir H. de la Beche, were simply folds and repetitions of the already classified Silurian rocks of Shropshire, Hereford, Radnor, etc. It is from this date that I considered my classification to be established on the broad European scale."

Resigning the chair to one of the founders of the Geo-logical Society, Henry Warburton, Murchison concluded his second and last tenure of the office. " I bid you farewell," he said to his fellow-members, " as friends in whose society, whilst acquiring knowledge, I have passed the happiest days of my life. . . . I have deeply felt the honour of presiding over men who in the course of a quarter of a century have demonstrated that there is no such thing as ' *odium geologi-cum*,' and whose members, rivals as they must be, have only sought to excel each other in their ardent search after truth."

Did the enthusiasm of the moment lead the writer to forget the very marked 'odium' which had been evoked during the early Devonian warfare ? Had the angry words of Macculloch vanished from his memory ? It was well, indeed, that they should, but not without leaving behind them just trace enough to keep him, even in the glow of excitement, from painting in too rosy a hue the intercourse of men whom even the brotherhood of science could not save from the ordinary frailties of humanity. To his eulogistic language the geological doings of after years furnished a comment of bitter irony, since his own name, to his deep grief indeed, and most unwillingly on his part, came to stand out prominently in the most noted instance of the *odium geologicum* which the history of British science has yet offered.

END OF VOL. I.

PRINTED BY T. AND A. CONSTABLE, PRINTERS TO HER MAJESTY, AT THE EDINBURGH UNIVERSITY PRESS.